Stochastic Space–Time Models and Limit Theorems

Mathematics and Its Applications

Stochastic Space–Time Models and Limit Theorems

edited by

L. Arnold

and

P. Kotelenez

Forschungsschwerpunkt Dynamische Systeme,
Universität Bremen, F.R.G.

D. Reidel Publishing Company

A MEMBER OF THE KLUWER ACADEMIC PUBLISHERS GROUP

Dordrecht / Boston / Lancaster

Library of Congress Cataloging in Publication Data
Main entry under title:

Stochastic space-time models and limit theorems.

    (Mathematics and its applications)
    Includes index.
    1. Stochastic analysis–Addresses, essays, lectures. 2. Stochastic
differential equations–Addresses, essays, lectures. 3. Limit theorems
(Probability theory)–Addresses, essays, lectures. 4. State-space methods–
Addresses, essays, lectures. I. Arnold, L. (Ludwig), 1937–
II. Kotelenez, P. (Peter), 1943– . III. Series: Mathematics and its
applications (D. Reidel Publishing Company)
QA274.2.S776   1985   519.2   85–10895

ISBN-13: 978-94-010-8879-4   e-ISBN-13: 978-94-009-5390-1
DOI: 10.1007/978-94-009-5390-1

Published by D. Reidel Publishing Company
P.O. Box 17, 3300 AA Dordrecht, Holland

Sold and distributed in the U.S.A. and Canada
by Kluwer Academic Publishers,
190 Old Derby Street, Hingham, MA 02043, U.S.A.

In all other countries, sold and distributed
by Kluwer Academic Publishers Group,
P.O. Box 322, 3300 AH Dordrecht, Holland

Table of Contents

Part II. Limit Theorems

SERIES EDITOR'S PREFACE

Approach your problems from
the right end and begin with
the answers. Then one day,
perhaps you will find the
final question.

'The Hermit Clad in Crane
Feathers' in R. van Gulik's
The Chinese Maze Murders.

It isn't that they can't see
the solution.
It is that they can't see the
problem.

G.K. Chesterton. The Scandal
of Father Brown 'The Point of
a Pin'.

Growing specialisation and diversification have brought a
host of monographs and textbooks on increasingly specialized
topics. However, the "tree" of knowledge of mathematics and
related fields does not grow only by putting forth new
branches. It also happens, quite often in fact, that
branches wich were thought to be completely disparate are
suddenly seen to be related.

Further, the kind and level of sophistication of mathematics
applied in various sciences has changed drastically in
recent years: measure theory is used (non-trivially) in
regional and theoretical economics; algebraic geometry
interacts with physics; the Minkowsky lemma, coding theory
and the structure of water meet one another in packing and
covering theory; quantum fields, crystal defects and
mathematical programming profit from homotopy theory; Lie
algebras are relevant to filtering; and prediction and
electrical engineering can use Stein spaces. And in
addition to this there are such new emerging subdisciplines
as "experimental mathematics", "CFD", "completely integrable
systems", "chaos, synergetics and large-scale order", which
are almost impossible to fit into the existing classification
schemes. They draw upon widely different sections of mathematics.
This programme, Mathematics and Its Applications, is devoted
to new emerging (sub)disciplines and to such (new)
interrelations as exampla gratia:

-a central concept which plays an important role in several
different mathematical and/or scientific specialized areas;
-new applications of the results and ideas from one area
of scientific endeavour into another;
-influences which the results, problems and concepts of one
field of enquiry have and have had on the development of
another.

The Mathematics and Its Applications programme tries to make
available a careful selection of books which fit the philosophy
outlined above. With such books, which are stimulating rather
than definitive, intriguing rather than encyclopaedic, we
hope to contribute something towards better communication
among the practitioners in diversified fields.

It's remarkable how fast sometimes a new coherent mathematical
subdiscipline can establish itself. One such area is concerned
with stochastic space-time models including pattern formations
and dynamically evolving patterns in space. This is a topic,
which as a recognized entity, did, perhaps even less than
a decade ago, not yet exist, and now the small part of it
which goes under the name of reaction-diffusion equations
has already at least two books published and another in
preparation.

As usual, these new fields tend not to fit very well in
existing classification schemes, making than a prime area
of concern of this book series.

It's also interesting to see how the evolution mathematical
thinking (and applications) repeats itself (more or less)
in patterns. In deterministic mathematics there are the lines
of development scalars $\rightarrow$ vectors $\rightarrow$ infinite dimensional spaces
and static $\rightarrow$ dynamic. And so, after stochastic variables,
stochastic processes, and stochastic differential equations
have established themselves as substantial fields of investigation, the time seems to have come for stochastic operators,
stochastic partial differential equations and stochastic
functional analysis.

There is much more involved than just generalizing everything
in sight to a stochastic context. There are deep fundamental
interactions between geometry (of the underlying spaces)
and stochastics on the one hand and stochastics and dynamics –
the main concern of the present volume – on the other. There
are lots of surprises, especially for such as myself who
were mainly trained deterministically.

That is what makes – in my view – the field so fascinating, as the thirteen state-of-the-art surveys in this volume will testify.

The unreasonable effectiveness of mathematics in science....

Eugene Wigner

Well, if you know of a better 'ole, go to it.

Bruce Bairnsfather

What is now proved was once only imagined.

William Blake

As long as algebra and geology proceeded along separate paths, their advance was slow and their applications limited.
But when these sciences joined company they drew from each other fresh vitality and thenceforward marched on at a rapid pace towards perfection.

Joseph Louis Lagrange

Bussum, March 1985   Michiel Hazewinkel

PREFACE

Stochastic space-time models describe phenomena which change with time, are distributed in space and contain fluctuations. Those models are of growing importance in various fields such as physics and chemistry (many particle systems, quantum field theory, reaction-diffusion models), biology (population dynamics) and engineering (random loads on mechanical structures). They also give rise to challenging mathematical problems. The mathematics of space-time phenomena, which is just about to emerge, is basically a combination of the modern theory of stochastic processes and functional analysis.

The present volume presents the state of the art in this new field. It contains 13 invited papers of a workshop held in November 1983 at the University of Bremen delivered by leading specialists in the field. The papers are preceded by a unifying introduction. The main subjects are:
(i)   Stochastic partial differential equations (existence, uniqueness and regularity of solutions),
(ii) Stochastic analysis and Markov processes in infinite dimensions,
(iii) Limit theorems, applications.

Finally, we want to thank the Reidel Publishing Company for their friendly cooperation and the University of Bremen for the financial support to our workshop.

P. Kotelenez

STOCHASTIC SPACE-TIME MODELS AND LIMIT
THEOREMS: AN INTRODUCTION

1. Stochastic Partial Differential Equations (SPDE's) as Stochastic Space-Time Models

Stochastic space-time models describe phenomena
which change with time, are inhomogeneous in space
and depend on chance. They arise in various fields
of physics, chemistry, biology and engineering.
Both analogy to the theory of spatially homogeneous
stochastic systems and other reasoning suggest that
a class of stochastic space-time phenomena could be
modelled as solutions of SPDE's which can be
formally written as a (deterministic) partial
differential equation (PDE) perturbed by a
"suitable" noise term

$$\frac{\partial}{\partial t} X(t,r,\omega) = P(t,r,X,\partial_i X) + \xi(t,r,X,\partial_i X,\omega)$$

(1)

(initial condition, boundary condition –
if necessary).

Here $r \in D$, where D is some open domain in
R^n and P is a PD operator, densely defined on
$L_2(D)$.

There are essentially two sources for the
stochasticity in (1), namely internal noise and
external noise.

In the case of internal noise the unperturbed
PDE ($\xi \equiv 0$ in (1)) is the phenomenological or macroscopic description of a system which consists of a
large number of subsystems, like the (average)
density of a chemical reactant in a reactor. Thus
(1) with $\xi \equiv 0$ "is only approximate, and in
reality there will be small deviations from them
which show up as fluctuations" (van Kampen [8]).
These fluctuations are modelled by $\xi(t,r,X,\partial_i X,\omega)$,
and (1) describes a special class of synergetic

1

L. Arnold and P. Kotelenez (eds.), Stochastic Space-Time Models and Limit Theorems, 1–9.
© *1985 by D. Reidel Publishing Company.*

systems in the sense of Haken [4] (cf. Section 3 for
a more detailed example as well as Nicolis and
Prigogine [13]).

External noise appears by exposing a deterministic
system to a random force. In engineering many
phenomena are of this type (cf. Krée for a general
analysis of such systems and Kotelenez, Section II,
Ex. 3, for an example from continuum mechanics).

2.   On the Solvability of SPDE's

By analogy to the spatially homogeneous (finite
dimensional) case one sets

$$dM(t,\cdot,X,\partial_i X,\omega) = \xi(t,\cdot,X,\partial_i X,\omega)dt$$

and interprets (1) as an infinite dimensional Itô
(evolution) equation with unbounded "drift" operator
P . Then one tries to apply one of the standard PDE
methods to (1), e.g., "variation of constants" if
P is linear and its closure generates a strongly
continuous semigroup, or the variational approach
if P is nonlinear and satisfies the variational
assumptions. In the case of a second order equation
one transforms this equation into a 2-system of
first order equations on a suitable product space
and obtains an evolution equation of type (1) with
unbounded drift.
Now let $\xi(t,r,X,\partial_i X,\omega) \equiv \xi(t,r,\omega)$ be the standard
white noise in t and r . Then M(t) is the
cylindrical Brownian motion on $L_2(D)$, which
defines for each t only a weak (finitely additive)
Gaussian distribution on the cylinder sets of $L_2(D)$
with characteristic functional $\exp(-\frac{t}{2}|\phi|^2)$, where
$|\cdot|$ is the standard Hilbert norm of $L_2(D)$ and
$\phi \in L_2(D)$ (cf. Kuo [11]). However, since M(t) is
not σ-additive on $L_2(D)$ we cannot treat (1) as
an Itô evolution equation on $L_2(D)$. On the other
hand, we know from the theory of abstract Wiener
spaces that M(t) defines a σ-additive measure on
some enlarged Banach (distribution) space $B \supset L_2(D)$
(Kuo [11], Itô [6]). Consequently, one could try to
extend P to an unbounded operator $\tilde{P}$, which
satisfies on B the deterministic PDE assumptions,
in order to give a meaning to (1) as an SPDE

(Itô evolution equation) on B . It is easy to see
that this is even for linear operators in general
not possible (sufficient conditions for extendibility of linear operators to certain distribution
spaces are given in DaPrato and Grisvard [2]).
Moreover, since there is no general extension of
the pointwise multiplication of continuous functions
to the distributions (Schwartz [14]), we cannot
extend a nonlinear P with polynomial part (cf.
the example of Section 3). But there may be a weak
dimension dependent solvability of (1) for certain
nonlinear P on bounded domains D and cylindrical
Brownian motion M(t) as an equation on $L_2(D)$
(without extending!), using the smoothening
property of the semigroup generated by the (linear)
main part of P (cf. Section 4 and Kotelenez). On
the other hand, Dawson [3] has obtained a diffusion
approximation to a system of branching Brownian
motions which is measure valued and satisfies an
equation of type (1) with state dependent noise
(driven by the cylindrical Brownian motion) only
formally, i.e., the operations in (1) are not
separately defined. In short, if M(t) does not
define a σ-additive measure on $L_2(D)$, the
treatment of (1) as an Itô evolution equation (in
the sense described above) excludes interesting
stochastic space-time models which formally satisfy
(1). Therefore, one may wonder whether we cannot
drop the axiom of σ-additivity (Kolmogorov [9]) and
solve (1) directly on $L_2(D)$ in the setting of
weak (finitely additive) measures (cf. Kallianpur
and Karandikar [7] for applications of weak measures
in nonlinear filtering). To our knowledge this is
still an open problem.

3.   A Key Example

Let us now discuss an example, where (1) arises in
a "natural" way from the unperturbed PDE. For the
sake of ease we assume D to be a bounded convex
domain with smooth boundary. Set

$$P(t,r,X,\partial_i X) = \Delta X + f(X), \tag{2}$$

where Δ is the Laplacian closed w.r.t. homogeneous

Dirichlet boundary conditions and f is a
polynomial with constant coefficients s.t.
$f(o) \geq 0$ and for nonlinear f the leading
coefficient is negative. Moreover, we assume a
positive initial condition $X(o,r) \geq 0$. Then, for
$\xi \equiv 0$, (1) becomes the classical reaction and
diffusion equation for the density of one reactant
in the reactor D , which has a positive solution
$X = X(t,r)$. The validity of this (macroscopic)
description rests on the implicit assumption that
the number of particles in D is large (~infinite) -
as indicated in Section 1. Consequently, we can try
to derive X as the thermodynamic limit of an
appropriately rescaled sequence of stochastic
(counting) processes for the density of the reactant,
if the number of particles tends to infinity. This
can be done as follows: Divide D into N cells
of volume v (at least in the interior) and define
a jump Markov process $X^{v,N}$ by counting the number
of particles in each cell and dividing this number
by v , where discretizations of Δ and $f(x)$ are
used to describe the random change in the number of
particles due to diffusion (between cells) and
reaction (within cells). Arnold and Theodosopulu
[1] have shown that for a certain sequence (v,N)
s.t. $vN \to \infty$, $N \to \infty$

$$X^{v,N} \longrightarrow X , \tag{3}$$

which is called the law of large numbers (LLN). The
next canonical step is to compute the fluctuations
around X by a central limit theorem (CLT), i.e.,
find a sequence of positive numbers $\alpha(v,N)$ s.t.

$$\alpha(v,N)(X^{v,N}-X) \Rightarrow Y , \tag{4}$$

where Y is some Ornstein-Uhlenbeck process. Y is
called the correction term to the deterministic
PDE for the density X , i.e., if (4) is correct,
we have $X^{v,N} \sim \dfrac{1}{\alpha(v,N)} Y$.
The third problem is to describe Y , if possible,
as the solution of a (linear) SPDE

$$dY(t) = AY(t)dt + dM(t), \tag{5}$$

where M(t) is some Gaussian martingale and A is
an unbounded "drift". If, e.g. the ratio $\dfrac{v}{N}$ is

chosen in such a way that M(t) is white noise in
the spatial direction or of the order of the
derivative of white noise then it is desirable to
obtain spatially smooth versions of Y .
The CLT (4) (and the LLN (3)) was obtained by
Kotelenez [10] under the assumption that the
reaction is linear, and it was shown that Y
satisfies (5) with M(t) of the order of the
derivative of white noise in the spatial direction
and $A = \tilde{\Delta} + f'$, where $\tilde{\Delta}$ is the extension of Δ
to appropriately chosen Hilbert distribution spaces.
The mathematical difficulty for getting a Gaussian
distribution valued Y as correction term in the
case of nonlinear reaction (which is an example
of local interaction of particles) is linked to the
problem of Section 2, namely that there is no
general extension of the pointwise multiplication
to the distributions. Moreover, "correcting" the
nonlinear PDE for the density X by white noise
seems to contradict physical observation. Consequently, physicists suggest a different scaling
for nonlinear reaction and diffusion systems near
bifurcation points (Van den Broeck).

4.   Contents of these Proceedings

The following problems emerge from the previous
considerations:
a) Describe stochastic space-time models as
(possibly distribution valued) stochastic processes
Y .
b) Derive Y as the limit of stochastic processes
on lattices or of positions of finitely many
particles etc., where we can easily define interaction through multiplication and other mathematical
operations. These problems were the major topics of
our workshop.

To a) Typically Y will live on a nuclear space
Φ' which is the strong dual of a nuclear space Φ
(of smooth elements) with a Hilbert space H (e.g.
$H = L_2(D)$) in between, where H is identified
with its dual H' ,

$$\Phi \subset H = H' \subset \Phi' . \tag{6}$$

If Φ is also a Fréchet space then it is the
projective limit of a sequence of Hilbert spaces
H_α, whence Y can be analyzed on a scale of
Hilbert spaces

$$\Phi \subset H_\alpha \subset H_O = H_O' \subset H_{-\alpha} \subset \Phi' , \qquad (7)$$

where $\alpha \in N$ (or R_+), $H_O := H$ and $H_{-\alpha} = H_\alpha'$ with
dense continuous imbeddings and the nuclearity
property. It has been shown in Section II of
Kotelenez that (1) makes sense on some $H_{-\alpha}$ for
$H_{-\alpha}$-valued state independent noise M(t) and linear
P by extending P. Moreover, if M(t) is state
dependent, (1) can be given a rigorous meaning if
the main part of P is linear and generates an
analytic semigroup U(t) s.t. U(t)M(t) $\in H_O$
(cf. Kotelenez). Thus, at least a certain class of
space-time models with distribution valued noise
input can be represented by SPDE's (interpreted as
Itô evolution equations) on some Hilbert space. The
papers of Ichikawa, DaPrato and Section I of
Kotelenez are concerned with properties of SPDE's
on a fixed abstract Hilbert space.

As explained in Section 2 the treatment of (1)
depends on the Itô integral in infinite dimensions.
This integral has been well described by Metivier
and Pellaumail [12] if the state space is a Hilbert
space. On the other hand, even if we can show
existence and uniqueness of (1) (as well as tightness of some approximation to (1)) on some Hilbert
space, this space may be not the most "natural"
space. An example is the Ornstein-Uhlenbeck process
Y obtained by Holley and Stroock [5] as the
Gaussian approximation to a system of branching
Brownian motions, which solves (5) on some Hilbert
distribution space but is ergodic on some Banach
distribution space. Thus, it is desirable to extend
the Itô integral to Banach spaces, which is done by
Dettweiler. Ustunel deals with generalizations of
basic stochastic concept to the nuclear triple (6),
including SPDE's with a finite dimensional Wiener
process as driving force. However, we have seen in
Section 2 that the concept of an SPDE of type (1)
may be too narrow for many nonlinear models. A
different approach to nonlinear stochastic space-

time models (linked to quantum field theory) is given by Albeverio, Høegh-Krohn and Holden. Sections I and II of their paper deal with the construction of Markov fields through Dirichlet forms on infinite dimensional spaces, Section III with the construction of Markov surfaces, i.e., of random fields which do not depend on points but on (n-1)-dimensional hypersurfaces and take values in a Lie group. Krée's paper deals with the description of SPDE's arising from external noise. The (generalized stationary) Gaussian driving term is obtained as the solution to the linear SPDE (5), where A generates an asymptotically stable linear semigroup. Moreover, smoothness properties of probability measures on infinite dimensional spaces are derived in the framework of Banach valued Sobolev spaces.

To b) In Rost's paper weak convergence of a system of interacting particles to a generalized Ornstein-Uhlenbeck process is proved in the set-up (7). Zessin gives a weak convergence criterion in the set-up (6) in terms of moments. Grigelionis and Mikulevičius derive weak convergence criteria for semimartingales with values in a rigged Hilbert space

$$B \subset H = H' \subset B' \qquad (8)$$

by extending finite dimensional results in terms of local characteristics to (8), where H is a Hilbert space identified with its dual, B is a Banach space densely and continuously imbedded into H and B' its dual. In Metivier a weak compactness criterion for Hilbert space valued martingales is given and Pardoux proves a wide-band noise approximation for a semi-linear SPDE. Finally, Van den Broeck deals with the problem of Gaussian and non-Gaussian approximation for nonlinear particle systems (cf. Section 3).

Peter Kotelenez
Forschungsschwerpunkt Dynamische Systeme
Universität Bremen
Bibliothekstraße
Postfach 330 440
2800 Bremen 33
West Germany

5.   References

[1] L. ARNOLD and M. THEODOSOPULU: "Deterministic
    Limit of the Stochastic Model of Chemical
    Reactions with Diffusion", Adv. Appl. Prob.
    12 (1980) 367 - 379

[2] G. DA PRATO and P. GRISVARD: "Maximal
    Regularity for Evolution Equations by Inter-
    polation and Extrapolation" to appear in
    J. functional Analysis

[3] D. A. DAWSON: "Stochastic Evolution Equations
    and Related Measure Processes", J. Multivar-
    iate   Anal. 5 (1975) 1 - 52

[4] H. HAKEN: "Advanced Synergetics", Springer-
    Verlag, Berlin-New York 1983

[5] R. HOLLEY and D. W. STROOCK: "Generalized
    Ornstein-Uhlenbeck Processes and Infinite
    Particle Branching Brownian Motions", Publ.
    RIMS, Kyoto Univ. 14 (1978), 741 - 788

[6] K. ITÔ: "Continuous Additive S'-Processes",
    in B. Grigelionis (ed.) "Stochastic Differen-
    tial Systems", Springer Verlag, Berlin-
    New York 1980

[7] G. KALLIANPUR, R. L. KARANDIKAR: "White Noise
    Calculus and Nonlinear Filtering Theory" to
    appear in Annals of Probab.

[8] N. G. VAN KAMPEN: "Stochastic Processes in
    Physics and Chemistry", North Holland,
    Amsterdam-New York 1983

[9] A. N. KOLMOGOROV: "Grundbegriffe der Wahr-
    scheinlichkeitsrechnung" Ergebn. d. Math. 2,
    Heft 3, Berlin 1933

[10] P. KOTELENEZ: "Law of Large Numbers and
     Central Limit Theorem for Chemical Reactions
     with Diffusion", Ph. D. Thesis, Bremen 1982

[11] H. H. KUO: "Gaussian Measures in Banach Spaces",
     Springer Verlag, Berlin-New York 1975

[12] M. METIVIER and J. PELLAUMAIL: "Stochastic
     Integration", Academic Press, New York-
     London 1980

[13] G. NICOLIS and I. PRIGOGINE: "Self-Organiza-
      tion in Non-equilibrium Systems", John Wiley
      & Sons, New York-London 1977

[14] L. SCHWARTZ: "Sur l'impossibilité de la
      multiplication des distribution", C. R. Acad.
      Sci., Paris 239 (1954), 847 - 848

S.Albeverio,R.Høegh-Krohn,H.Holden

MARKOV PROCESSES ON INFINITE DIMENSIONAL SPACES,
MARKOV FIELDS AND MARKOV COSURFACES

ABSTRACT: We review work on Dirichlet forms and symmetric
Markov processes on infinite dimensional spaces. Especially
we consider the connections with the construction of
homogeneous generalized Markov random fields. We also discuss
a non commutative extension to the case where the state
space is a group. The extension involves a stochastic
calculus for group valued mappings defined on hypersurfaces
of codimension 1.

1. INTRODUCTION

    The theory of Dirichlet forms and of symmetric Markov
processes was developed as a powerful L^2-extension of
classical potential theory and the associated processes.
The basic theory is beautifully presented in [1],[2],
[3] (for newer developments see also [4]). An ideal field
of application of the theory is quantum mechanics, where
the L^2-extension of classical C-results is needed, both
because the natural space to work with is a Hilbert space
and because the drift coefficients and potential functions
involved in important examples have singularities of various
sorts. In particular it was soon realized that the usual
existence and uniqueness results of stochastic differential
equations can not be applied to the stochastic equations
arising in quantum mechanical situations (via the stochastic
mechanical formulation of quantum mechanics [5], [6]).
A stochastic calculus based on the theory of Dirichlet forms
was developed, simultaneously, for the needs of quantum
mechanics, by Albeverio, Høegh-Krohn and Streit in [7], [8]
and in generality by Fukushima in [1], [9], [10]. Other
examples of a close interaction between quantum mechanical
problems and the theory of Dirichlet forms are provided by
the study of criteria for closability of Dirichlet forms
[1], [8], [11], [12], [13], [14], the problem of approximation
of Dirichlet forms and associated processes [13], [15],

11

L. Arnold and P. Kotelenez (eds.), Stochastic Space-Time Models and Limit Theorems, 11–40.
© *1985 by D. Reidel Publishing Company.*

[16], [71], the problem of uniqueness of Dirichlet forms and
associated self-adjoint operators [1], [7], [8], [17], [18],
[19], [71], the problem of the relation between Dirichlet
forms and Schrödinger operators [6-8], [16], [20-25], the
problem of ergodicity and barriers for the processes [4],
[5], [6], [26-32], the problem of densities for the processes
[3][4]and many other.

     For some applications of the theory of Dirichlet forms
outside quantum mechanics see e.g. [1-3], [14], [23],
[34-42]and references therein.
The theory of Dirichlet forms required for all these
applications is the basic one in which the underlying space
is locally compact. For applications to quantum field
theory is was realized quite early [7] that an extension
to the case of an underlying infinite dimensional linear
space (e.g. a Hilbert space) is needed. It can actually
be shown [43] that every local Dirichlet form can be looked
upon as a Dirichlet form associated with some Hilbert space,
so that the theory of Dirichlet forms associated with
Hilbert spaces properly extends the usual one on locally
compact spaces, at least in the local (diffusion) case.
The construction of a Dirichlet form and associated processes
on infinite dimensional spaces was begun by Albeverio and
Høegh-Krohn in [7], subsequently worked by the same authors
in [44], [45] (see also [23], [46] for applications) and
by Paclet [47], [48] and Kusuoka [37] (see also [49] for
a recent application).
More recently Albeverio, Fenstad, Høegh-Krohn and Lindstrøm
[50] have given a version of the theory of Dirichlet forms
in terms of non standard analysis, which unifies the finite
dimensional and the infinite dimensional theory.

     In Sect. I of this paper we discuss some basic proper-
ties and some new results on Dirichlet forms, both in the
finite dimensional and in the infinite dimensional case.
The study of Dirichlet forms is actually equivalent with
the study of symmetric Markov processes. In the case where
the state space is an infinite dimensional function or
distributional space it appear quite natural to try to look
at the processes as random fields or generalized random
fields. An especially interesting class would be the one in
which the (generalized) random fields have themselves a
Markov property and are homogeneous (which extends the
stationarity and symmetry properties of the processes). It

was realized by Nelson, see e.g. [51], and in [7] that
quantum fields provide, at least in principle, natural
examples of such fields. Technically, the construction of
global homogeneous Markov fields has turned out to be very
difficult, but by now several examples are known [52-54]
(see also [56] for expositions of the method introduced by
Albeverio and Høegh-Krohn in [52]).
The construction involves, among other things, the solution
of a stochastic Dirichlet problem, with distributional data,
which have been recently treated in great generality with
powerful methods of axiomatic potential theory [57-60]
(see also [61] for related methods).
In Sect. II of this paper we shall discuss briefly these
connections.
There are several directions in which extensions of the ideas
of Sect. I and II can be looked for, and again the
inspiration has come from problems of quantum theory (quantum
gauge fields). The case where the underlying space has the
richer structure of a manifold has been discussed by
Albeverio and Høegh-Krohn in [23].
Extensions to the cases where the state space is a group
have been obtained following two methods. One inspired by
the theory of group representations, providing algebraic
analogues of the random fields. This has lead to a series
of papers [62], [63](and references therein), which have
been partly reviewed elsewhere [23], so that this method
will not be discussed here.
Another method, which is presented here, is the one in which
the pointwise defined random field is replayed by a random
field associated with d-1-dimensional hypersurfaces in a
d-dimensional manifold, the value being again in a group.
This approach leads in particular to the development of a
theory of stochastic integrals for d-1-forms which extends
the theory of Brownian motion on a Lie group, and of the
associated integrals, to the case of multidimensional time
parameters.
Sect. III in the present paper reviews some of the recent
developments in this area.

I. Dirichlet forms and symmetric Markov processes on
   infinite dimensional spaces

In quantum field theory the energy operator
(Hamiltonian) is formally given by an analogue of the
classical Dirichlet forms, namely by

$$\frac{1}{2} \int (\nabla f)^2 d\nu ,$$

with ν a certain measure on an infinite dimensional linear
space (the so called "time zero vacuum measure") and f any
smooth cylinder function, ∇f being then a naturally defined
gradient of f. The mathematical realization of this idea has
been given by Albeverio and Høegh-Krohn in [44] (see also
[7], [45]). The study of Dirichlet forms of the above type
(and of related ones) has been undertaken also from other
points of view by S. Kusuoka [37] and Ph. Paclet [47], [48],
see also [50] for a recent formulation in terms of non-
standard analysis.
Let us here shortly illustrate the main ideas of the approach
[44], recalling at the same time the corresponding problems
in finite dimensions.

In the finite dimensional situation ν is a Radon measure
with support $\mathbb{R}^d$ (but the general theory holds also for
locally compact Hausdorff space, with countable base for the
topology). A Dirichlet form E on $L^2(\nu)$ is then a bilinear
symmetric, nonnegative, closed form, with dense domain
$D(E) \subset L^2(\nu)$ s.t. $E(f^\#, f^\#) \leq E(f,f)$ for $f^\# = (f \vee 0) \wedge 1$, for
all $f \in D(E)$ (the fact that E "contracts" under $f \to f^\#$ is the
"Dirichlet property"). The importance of such forms is that
they are in 1 - 1 correspondence with (ν-) symmetric Markov
semigroups $P_t = e^{-tH}$, $t \geq 0$, in $L^2(\nu)$, with H self-adjoint,
nonnegative, where Markov means $0 \leq f \leq 1 \to 0 \leq P_t f \leq 1$, $\forall f \in L^2(\nu)$
and ν-symmetric means that the adjoint P_t^* of P_t in $L^2(\nu)$
coincides with P_t itself. The correspondence $E \leftrightarrow (P_t, t \geq 0)$
is given by $E(f,f) = (H^{1/2}f, H^{1/2}f)_{L^2(\nu)}$, with
$(,)_{L^2(\nu)}$ the scalar product in $L^2(\nu)$. Symmetric Markov
semigroups give rise to symmetric Markov processes X_t^ν with

invariant measure ν.
If E is a regular Dirichlet form then to it there is
naturally associated a Hunt process. This process is a
diffusion process (in the sense of continuous paths) iff
the regular Dirichlet form is local.
Of course there are subleties in this general results, but
we can here only refer to the literature for more details.

In quantum mechanics natural Dirichlet forms are given
first on $C_o^\infty(\mathbb{R}^d)$ by $E_o(f,f) = \frac{1}{2} \int_{\mathbb{R}^d} (\nabla f)^2 d\nu$, with

$\nabla = (\frac{\partial}{\partial x^1} , \ldots, \frac{\partial}{\partial x^d})$ the gradient operator.

1) For which Radon measures ν does E_o have a closed
extension? Is it then indeed a Dirichlet form? A regular one?
A local one? Criteria for ν to yield closed extensions of E_o
are known in the literature, and have been reviewed in [11]
(see also [14]). They are basically of 2 types, requiring
ν to have a density with respect to Lebesgue measure
$\lambda(dx) = dx$ (this is also necessary for d=1 [12]) and in

addition either $\rho > 0$ λ-a.e., $\frac{\partial}{\partial x^i} \rho^{1/2} \in L^2_{loc}(U)$, $U \subset \mathbb{R}^d$

open, $\nu(\mathbb{R}^d-U) = 0$ or else $\rho > 0$ on compacts,

$\rho \in L^1_{loc}(U)$, $U \subset \mathbb{R}^d$ open, s.t. $\lambda(\mathbb{R}^d-U) = 0$.
Regularity holds always for the closure of E_o, wherever it
exists, but might fail for other closed extensions. Locality
always holds for the closure (not so for other extensions
[1]).

2) Is there some uniqueness, so that all closed
extensions of E_o coincide? This has been proven e.g. for
$\rho > 0$, locally Lipschitz by N. Wielens [19].
One can prove that this is equivalent with the self-adjoint
operator $\mathcal{L}$ associated with the closure of E_o to be already
essentially self-adjoint on $C_o^\infty(\mathbb{R}^d)$. The weaker property of
"Markovian uniqueness", to the extent that all self-adjoint
extensions of $\mathcal{L}$ leading to Markov semigroups (Dirichlet
forms) coincide has been discussed in [17] and [19].

Two important closed extensions of E_o are the "minimal" one E, the closure of E_o, and the "maximal one" $\widetilde{E}$ with domain $D(\widetilde{E}) = \{f \in L^2(\nu) \mid \nabla f \in L^2(\nu)\}$.

3) Study of the symmetric semigroup associated with the Dirichlet form, construction of a process associated with it, construction of a "regular" realization (diffusion process), study of the corresponding potential theory. In the finite dimensional case these question have been studied to a great extent, we refer to [1-3] for basic references and [4], [7], [8], [13], [15], [16-25], [27-30], [33], [40-46], [50], [56-61], [64], [71] for some applications to quantum mechanics and other domains of physics.

4) Questions of ergodicity of P_t or of ν with respect to space translations have been discussed mathematically [6-8], [27-32], [44], [45], also for their direct inter-pretation in physical terms. In particular connections between capacity zero sets, nodes of ρ and tunneling – non tunneling across the nodes of ρ have been discussed, as well as connections between capacity and scattering theory for the associated Hamiltonians (cfr. 5)). Recently this has also acquired new actuality through the discussion of the method of stochastic quantization (propagated particularly by Parisi and Wu).

5) Correspondence between P_t and the "Hamiltonian semi-group" T_t in $L^2(\mathbb{R}^d, \lambda)$ given by $T_t = \varphi^{-1} P_t \varphi$ if $\rho = \varphi^2$, $\varphi > 0$ λ-a.e. this has been discussed in [7], [8][24], [25]; a very recent reference on this topics with further references is [25].

6) Approximation of P_t, the corresponding process, and of T_t in 5) by corresponding objects constructed with approximations φ_n of φ, with suitable φ_n (so that one has a good control on the corresponding P_t^n, T_t^n constructed from φ_n in the way as P_t, T_t was constructed from φ). For these problems see [15], [13] and the recent paper [16].

We shall now shortly mention few results on corresponding problems for the infinite dimensional situation.

1) Similarly as in the finite dimensional situation the starting point is the densely defined, not yet closed, form $\frac{1}{2} \int (\nabla f)^2 d\nu$ in $L^2(\nu)$. ν is here a Radon probability measure on some infinite dimensional linear space Q', realizing in some sense $\mathbb{R}^d$ for $d = \infty$. The choice we made (in [7], [44], [45]) is Q' the dual of a countably normed nuclear space Q such that $Q \subset Q'$ and such that there exists a real separable Hilbert space $\mathcal{H}$ with $Q \hookrightarrow \mathcal{H} \hookrightarrow Q'$, with continuous injections $\hookrightarrow$. $\mathcal{H}$ is used to define ∇ in as much as ∇ is defined as a map from the space FC^∞ of smooth function on Q' with base in finite dimensional subspaces of H spanned by elements in Q with values in $L^2(\nu) \times \mathcal{H}$ (in finite dimensions ∇ can correspondingly be looked upon as a map from $C^\infty(\mathbb{R}^d)$ into $L^2(\nu) \hat{\otimes} \mathbb{R}^d$). Let then $E_o(f,f) = \frac{1}{2} \int (\nabla f \cdot \nabla f)_{\mathcal{H}} d\nu$ for $f \in FC^\infty$, with $(,)_{\mathcal{H}}$ the scalar product in $\mathcal{H}$. In [7], [40], [41] we defined as a realization of a "Dirichlet form extending E_o" as the closure of E_o in $L^2(\nu)$, whenever the closure exists. Sufficient conditions for ν such that the closure exists have been given in [45] (which extends slightly the results of [7], [44]) and correspond to conditions of the type $\frac{\partial}{\partial x^i} \rho^{1/2} \in L^2(\mathbb{R}^d)$ in finite dimensions, namely that $1 \in D(\nabla_i^*)$ with ∇_i the directional derivative in the direction of the basis vector $e_i \in Q$ in $\mathcal{H}$, where D means domain. Other realizations of Dirichlet forms in infinite dimensions are given by Kusuoka in [37]. Here basically Q' is replaced by a separable Banach space B and Q by the dual of B. The construction of E is replaced by $\tilde{E}(f,f) = \frac{1}{2} \int (Df,Df)_{\mathcal{H}} d\nu$ with domain $D(\tilde{E}) = \{f \in L^2(\nu) \mid \int (Df,Df)_{\mathcal{H}} d\nu < \infty$. Df is defined for any real measurable f on B which is both "ray absolutely continuous" (in the sense that for all $q \in B$ there exists a real measurable $\tilde{f}_q$ on B s.t. $f = \tilde{f}_q$ ν-a.s. and

$f(\cdot + tq)$ is absolutely continuous in $t \in \mathbb{R}$) and "stochastic
Gâteaux differentiable with respect to ν" (in the sense that
$\frac{1}{t}[f(\cdot+tq)-f(\cdot)]$ converges as $t \downarrow 0$, ν-stochastically).
One sets $(Df,q) = \lim_{t\downarrow 0} \frac{1}{t}[f(\cdot+tq)-f(\cdot)]$ (so that (Df,q) is
the stochastic Gâteaux derivative of f in the direction q).
Under point 2 we shall discuss a little the relation between
$\tilde{E}$ and the above E. Yet another realization of Dirichlet forms
in infinite dimensions has been given by Paclet in [47], [48].
It uses Q,Q' both separable real Hilbert space, s.t. $\mathcal{H} \subset Q'$,
$Q \subset \mathcal{H}$ with Hilbert-Schmidt injections i resp. i', the
transpose of i (after identification of $\mathcal{H}$ with its dual).
∇ is defined in a natural way on the space $\mathcal{H} C_b^1(Q')$ (of universally Lusin measurable [65], [66]) bounded functions f
s.t. for all $\xi \in Q'$, $h \to f_\xi(h)$ is Fréchet C^1 on $\mathcal{H}$ and
$\xi \to \nabla f_\xi(0)$ is bounded universally Lusin-measurable), with
domain $\mathring{D}(\nabla) = \{f \in L^2(\nu) \mid \nabla f \in L^2(\nu)\}$. Both Kusuoka and Paclet
thus operate rather with an infinite dimensional extension
of the concept of maximal Dirichlet form $\tilde{E}$ in finite dimension
(cfr. point 2 above), rather than of the minimal one. Both
in Kusuoka and Paclet the closedness of the forms is a consequence of a strict positivity condition of ν.

   2) Correspondingly as in the finite dimensional case the
question arises of the relation between the different Dirichlet
forms. This is as of yet less clarified than in finite
dimensions. A strong uniqueness result (essential self-
adjointness of the self-adjoint operator associated on E,
restricted on FC^∞) is proven in [7], [44] for the case where
ν is Gaussian (this includes the case where ν is the "time
zero measure" for the free quantum fields).
Recently Takeda [49] has proven a result concerning "Markovian
uniqueness", for the case where $\mathcal{H} = L^2(\mathbb{R})$, ν is absolutely
continuous with respect to Wiener measure (on $B = C_o[0,1]$).

The problem of uniqueness is a very important one concerning
the construction of canonical models of quantum fields, see
below.

   3) In all cases discussed above where Dirichlet forms
have been constructed one gets from them Markov symmetric
semigroups P_t in $L^2(\nu)$ (this is actually the main purpose
of constructing Dirichlet forms). From these one can always

get symmetric Markov processes with invariant measure ν.
The construction of nice realizations of the processes
("diffusions", weak solutions of the stochastic differential
equation d X_t = $\beta(X_t)$dt + dw_t, with dw_t the standard
Brownian motion associated with $Q \subset \mathcal{H} \subset Q'$ and $\beta(q)(\xi)$ for
$q \in Q$, $\xi \in Q'$ s.t. $\beta(q)(\xi) = -(q \cdot \nabla)^* 1(\xi))$ is more complex.
It has been undertaken in [45], in the framework of Albeverio,
Høegh-Krohn, and in [37], in the framework of Kusuoka (under
the assumption that the injection $\mathcal{H} \subsetneq B$ is compact and B is
either a Hilbert space or a Banach space with some additional
assumptions). In particular Kusuoka constructs a diffusion
process on the space B.

Potential theory can be discussed in all frameworks, in
particular [47] contains a detailed discussion of capacibility
of analytic sets.

4) Ergodicity questions have been discussed in the framework
of Albeverio - Høegh-Krohn in [7], [44]. These questions
should be relevant in connection with Parisi-Wu's approach
to stochastic quantization for quantum fields, but the
detailed applications have not yet been worked out. The
problem of nodes for the finite dimensional densities of ν
and its relations with questions of barriers for the corresponding process and quantum field theoretical tunneling
could probably be discussed in the framework of [7], [44],
[45].

5) Since in infinite dimensions there is no natural
Lebesgue reference measure, the semigroup T_t is a purely
formal object (but yet suggestive, e.g. in problems of
quantum fields, actually non standard analysis gives a
meaning to it, see [50]). In applications to quantum fields,
one takes $L^2(\nu)$ as the physical Hilbert space and the
generator H of $P_t = e^{-tH}$ as the Hamiltonian.
Smoothness property of the energy eigenfunction are statements about eigenfunctions of H, in particular the finite
dimensional elliptic result that the lowest end of the
spectrum consists of an isolated eigenvalue with strictly
positive eigenfunction corresponds here to the statement
that the measure ν has a strictly positive smooth density
with respect to finite dimensional subspaces, in a sense
made precise in [44]. It has been proven to hold in models of

quantum fields in [44].

    6) The approximation problem is a hard one already in
finite dimensions, and has not been studied in details in
the infinite dimensional case (it should here be thought of as
the problem of approximating the process associated with P_t
by "finite dimensional" processes); some information is
of course already available from the construction of the
Dirichlet forms and the corresponding semigroup and can be
extracted from [7], [44], [37], [47,48].

An interesting class of examples is furnished by quantum
field theory.

    A) Let us first consider the case of the so called "free
quantum fields". In this case $\mathcal{H} = L^2(\mathbb{R}^s)$, with s the
dimensionality of space (for the physical space we have s=3).
Q can be taken to be $\mathscr{S}(\mathbb{R}^s)$ (real Schwartz functions),
$Q' = \mathscr{S}'(\mathbb{R}^s)$ (tempered distributions).
ν is the standard Gauss measure with mean zero and covariance
$(-\Delta+m^2)^{-1/2}$, m some positive constant. In this case the
associated stochastic differential equation is
$dX_t = \beta(X_t)dt + dw_t$, with w_t Brownian motion associated with
$(Q,\mathcal{H},Q')$ (just the process with mean zero and increments
such $E(e^{i<q,(w_t-w_s)>}) = \exp(-\frac{1}{2} ||q||^2), \forall q \in Q$, with $|| \quad ||$
the norm in $L^2(\mathbb{R}^s)$, $<q,q'>$ the pairing of $q \in Q$, $q' \in Q'$) and
$\beta(q)(\xi) = - <\sqrt{-\Delta+m^2} q,\xi >$, for all $q \in Q$, $\xi \in Q'$).
Let E the Dirichlet form associated with ν as discussed in
1) and $P_t = e^{-tH}$, t > 0 the corresponding semigroup. Then
$H \upharpoonright FC_b^\infty(Q') = (-\frac{1}{2} \Delta - \beta \cdot \nabla) \upharpoonright FC^\infty(Q')$, with $\Delta \equiv \Sigma (e_i \cdot \nabla)^2$,
$\beta \cdot \nabla \equiv \Sigma \beta(e_i)(\cdot)(e_i \cdot \nabla)$, is essentially self-adjoint in $L^2(\nu)$,
as proven in [7], [44] (the proof only uses that ν is
Gaussian and $\beta \in L^2(\nu)$ i.e. $(q \cdot \nabla)^* 1 \in L^2(\nu)$ for any basis
vector $q \in Q$ in $\mathcal{H}$; it works for any Gaussian measures with
covariance A^{-1} s.t. A is trace class).
Thus in particular all Dirichlet forms extending E_o coincide.

There is then only one contraction semigroup in $L^2(\nu)$ with generator coinciding on FC^∞ with $H \upharpoonright FC^\infty$. The Ornstein-Uhlenbeck process X_t has nice realizations as a diffusion process, by the results of [7] and [44]. It is ergodic, zero being a simple isolated eigenvalue of H. ν has strictly positive and smooth density with respect to finite dimensional subspaces.

   B) The second class of examples we shall consider is the one given by interacting quantum fields in one space dimension (s=1).
Let $\mathscr{K}$,Q,Q' as in A) and let $\nu = \nu_\ell$, $0 \le \ell \le \infty$, where ν_ℓ is the "ground state measure" of model of quantum fields in one space dimension, with interaction in $(-\ell,+\ell)$. For $\ell < \infty$ one speaks of "space cut-off interaction", ν_ℓ is defined, as described in [44], by $\nu_\ell = \Omega_\ell^2 d\nu_0$, with ν_0 the measure as in A) and $\Omega_\ell \in L^1(Q',\nu_0)$ s.t. $\exists\ E_\ell \in \mathbb{R}$ with $(H_0+V_\ell)\Omega_\ell = E_\ell \Omega_\ell$, $E_0 = \inf\ \sigma(H_0+V_\ell)$, where $\sigma(\cdot)$ means spectrum and V_ℓ certain function on Q', $V_\ell \in L^p(Q',\nu_0)$ for all $1 \le p < \infty$ (additive in ℓ). V_ℓ is the space cut-off interaction. For $\ell = \infty$, ν_ℓ is defined as the restriction to the σ-algebra "associated with time zero fields" of the "Euclidean measure" describing a Euclidean quantum field theory in 2 space time-dimensions, see II. In a large class of models [44] one can show ν_ℓ, $1 \le \ell \le \infty$ yields a Dirichlet form E, hence by the construction of 3),a symmetric Markov semigroup and a corresponding diffusion process. ν_ℓ is smooth strictly positive ergodic, as shown in [44], which also contains further information on this situation. We shall add a few words on these models in Sect. II.

   C) The models considered by Parisi and Wu and others, in the so called method of "stochastic quantization", at least with regularizations, can be treated as processes associated with Dirichlet forms associated with $\mathscr{S}(\mathbb{R}^{s+1}) \subset L^2(\mathbb{R}^{s+1}) \subset \mathscr{S}'(\mathbb{R}^{s+1})$ of the general type discussed here, however with a different kind of ν than the fixed time Euclidean measures discussed in above examples A), B). It would certainly be worthwhile to carry through in details these constructions in the above framework.

II. The relation of the Dirichlet forms and processes of
    Sect. I with homogeneous Markov random fields.

    The theory in Sect. I was, but for the discussion of
examples connected with quantum fields, general, in
particular $Q, \mathcal{H}$, Q' did not have to consist of functions or
distributions in $\mathcal{S}'(\mathbb{R}^s)$. In this section however we shall
look upon the processes (X_t, ν) of Sect. I as "marginal
processes" obtained from a random field (X, μ) on $\mathbb{R}^d$, $d = s+1$.
Interesting examples are furnished by the generalized random
fields of Euclidean quantum field theory. However the
construction is more general.
Let μ be a measure on $\mathcal{S}'(\mathbb{R}^d)$ (or more generally some space
of functions or distributions over $\mathbb{R}^d$), with a support such
that for $\xi \in \text{supp } \mu$ one can define $<\xi, \rho>$ as an element in
$L^2(\nu)$, for all Borel measures ρ on $\mathbb{R}^d$ s.t.

$\int d\rho(x) d\rho(y) (-\Delta+m^2)^{-1} < \infty$ for some $0 \le m$ (m > 0 for d=1,2).

(Such measures exist: e.g. μ the Gaussian measure with mean
zero and covariance $(-\Delta+m^2)^{-1}$). Let for any closed measurable
$\Lambda \subset \mathbb{R}^d$, $\sigma(\Lambda) = \sigma(<\xi, \rho> \in L^2(\nu)$, supp $\nu \subset \Lambda)$. In particular we
have for $\rho(x) = \delta_t(x^1)\varphi(x^2, \ldots, x^d)$ that $<\xi, \rho> \in L^2(\nu)$ and
$\sigma(C_t)$ with $C_t = \{x \in \mathbb{R}^d, x^1 = t\}$ is well defined. Let
$\nu_C = \mu \upharpoonright \sigma(C)$. We assume there exists a structure $(Q, \mathcal{H}, Q')$
as in Sect. I s.t. ν_C is a probability measure on Q' and such
that $E_o(f,f) = \frac{1}{2} \int (\nabla f, \nabla f)_H d\nu_C$ in $L^2(\nu_C)$, $f \in FC^\infty$, is
closable. Then as in Sect. I we get by closure the minimal
Dirichlet form E(f,f) and the corresponding semigroup
$P_t = e^{-tH}$. Suppose μ is invariant under the transformations
U_t of $\mathcal{S}'(\mathbb{R}^d)$: $(U_t\xi) (x_1, \ldots, x^d) = \xi(x^1-s, x^2, \ldots, x^d))$
induced by the translations τ_t: $x \to (x^1+s, x^2, \ldots, x^d)$ in $\mathbb{R}^d$
along the axis orthogonal to C_t, $s \in \mathbb{R}$. Suppose also μ has
the global Markov property with respect to C_t, for all
$t \in \mathbb{R}$, in the sense that $E(f_+ f_- | \sigma(C_t)) = E(f_+ | \sigma(C_t)) E(f_- | \sigma(C_t))$,
$E(,) | \sigma(C_t))$ meaning conditional expectation, for all $f_{\pm}$
bounded and $\sigma(\{x^1 > t\})$ resp. $\sigma(\{x^1 < t\})-$

measurable. In this case one says that the generalized field (μ,X), with $X_\varphi(\xi) = <\xi,\varphi>$, $\varphi \in \mathscr{S}(\mathbb{R}^d)$, has the global Markov property with respect to half-planes [51].

We have that $E(.|C_0)U_t E(.|C_0) \upharpoonright L^2(\nu_{C_0})$ is a Markov semi-group P_t on $L^2(\nu_{C_0})$, with non negative infinitesimal generator H.

The semigroup is symmetric iff μ is invariant under reflections, in the sense of being invariant with respect to the transformation R induced on $\mathscr{S}'(\mathbb{R}^d)$ by $\rho: x \rightarrow (-x^1,x^2,\ldots,)$ on $\mathbb{R}^d$.

In the latter case H is self-adjoint and

$$(H^{1/2}f,H^{1/2}f)_{L^2(\nu_{C_0})} = \frac{1}{2} \int (\nabla f,\nabla f)_H d\nu_{C_0}$$

for all $f \in FC^\infty$.

H is thus a self-adjoint extension in $L^2(\nu_{C_0})$ of

$$-\frac{1}{2} \Delta - \beta - \nabla \upharpoonright FC^\infty, \text{ with } \beta(q)(q') = -(q,\nabla)^*1(q'), q \in Q, q' \in Q'.$$

In particular ν_{C_0} gives rise to a closable E_0, in the notation of Sect. I, and the Dirichlet form $(H^{1/2},H^{1/2})_{L^2(\nu_{C_0})}$ is a closed extension of E_0.

This then gives an interesting connection between random fields on $\mathscr{S}'(\mathbb{R}^d)$ (translation and reflection invariant) and symmetric Markov processes on Q'. In [44] we discussed a large class of examples where μ_ℓ is the Euclidean measure of quantum fields over $\mathbb{R}^2$ (d=2) with space cut-off so that the Euclidean interaction is in $\mathbb{R} \times (-\ell,+\ell) \subset \mathbb{R}^2$, $0 \le \ell \le \infty$, in which all hypothesis are satisfied, with $\mathscr{H} = L^2(\mathbb{R})$, $Q = \mathscr{S}(\mathbb{R})$, $Q' = \mathscr{S}'(\mathbb{R})$.

For $\ell < \infty$ all models which have been discussed in the literature have the required global Markov property. For $\ell = \infty$ by now 2 large classes of globally Markov Euclidean fields are known, those with trigonometric interactions [52] and those with exponential interactions [53], [54].

These are then Markov fields with marginals (in the above sense) which are symmetric Markov processes associated with Dirichlet forms of the local type, reducing to the above basic energy form $E_o(f,f)$ on smooth cylinder functions.

There is however a big open problem in this area, namely to prove uniqueness results for the associated semi-groups, of the type of problem 2) discussed above in Sect. I and II. The only results up to now, in the infinite dimensional case, are those of Takeda [49], who however treats a mathematical example with ν_{C_o} replaced by a measure absolutely continuous with respect to Wiener measure for Brownian motion.

In the proof of the global Markov property of the Euclidean fields [52] - [54] a solution of the Dirichlet problem with distributional data is involved. Recently a very general treatment of this problem has been given by Röckner, using among other techniques modern tools of axiomatic potential theory. The discussion on Gibbs states, initiated in [52], has also been extended considerably, to a theory of specifications in the sense of Preston.

Let us also remark that the situation discussed above with μ the Euclidean global Markov measure of quantum fields over $\mathbb{R}^2$ has other fascinating features, due to its full invariance with respect to all transformations induced on $\mathscr{S}'(\mathbb{R}^d)$ from Euclidean transformations in $\mathbb{R}^d$. In fact it gives rise to a canonical relativistic theory, as discussed in [21], [23], [46]. This means in particular that there is a representation of the Lorentz group in $L^2(\nu_{C_o})$, with infinitesimal generator of Lorentz transformation also given by a Dirichlet form, at least on FC^∞. There is an operator, canonically conjugate to the time zero field operator $X_\varphi(q') = \langle\varphi,q'\rangle$, $\varphi \in Q = \mathscr{S}(\mathbb{R}^s)$, $q' \in \mathscr{S}'(\mathbb{R}^s)$, in the sense of quantum mechanics (Weyl canonical commutation relation) and one has the "correct equation of motion".

We close this section by mentioning that there is another area in which Dirichlet forms on finite dimensional spaces have already played a role and by the relation through associated random fields the infinite dimensional Dirichlet forms can come to play a role. This is the area of unitary irreducible representations of "gauge groups" i.e. groups of mappings from a manifold into a Lie group. For work in this direction see [21], [23], [62]-[64].

III Markov cosurfaces – a more dimensional analogue of
   Brownian motion with values in a group

   In Sect. I we studied symmetric Markov processes with
values in infinite dimensional spaces and in Sect. II we
established their connection with Markov random fields. In
as much as Markov random fields are extensions from the case
of one dimensional time to the case of multidimensional time
the consideration of stochastic Markovian mappings from $\mathbb{R}^d$
(or a manifold) into a group G can be looked upon as a non
commutative extension of Markov random fields and the
associated Markov processes with infinite dimensional state
space. One such extension has been studied in details in
[21], [23], [62] – [64] It uses essentially methods of the theory
of representations of topological groups. In this lecture
we shall not discuss this extension, which has already been
reviewed elsewhere [23], [62]. We shall rather concentrate
on a new extension which is actually framed in terms of
random Markovian maps from hypersurfaces to Lie groups
("cosurfaces").

   For simplicity we shall first illustrate the concepts
in the discrete case in which $\mathbb{R}^d$ is replaced by a lattice
$\delta \mathbb{Z}^d \equiv M_\delta$, for some $\delta > 0$. We shall call <u>cell</u> γ of M_δ any
hypercube $\{x \in \mathbb{R}^d | n_i \delta \leq x_i \leq (n_i + 1)\delta, \ i=1,2,\ldots,d\}$, for some
$n_i \in \mathbb{Z}$.
We call <u>face</u> of such a cell γ of M_δ any hypersurface
$\{x \in M_\delta | x_i = n_i \delta\}$ or $\{x \in M_\delta | x_i = (n_i + 1)\delta\}$ for some i.
Clearly γ has 2d faces, exactly two orthogonal to any
given basic vector of $\mathbb{R}^d$. We shall consider oriented cells,
the orientation of a given cell being fixed by the one of
the <u>basic</u> cell $\{x \in \mathbb{R}^d | 0 \leq x_i \leq \delta, \ i=1,\ldots,d\}$, which is the
one given by the basic vectors $e_1,\ldots,e_d$ in $\mathbb{R}^d$. A face F
belonging to a given cell γ, so that $F \subset \partial \gamma$, has then a
corresponding orientation.

   We shall call H_δ the family of all finite unions of
oriented faces of cells of M_δ.
For $S_1, S_2 \in H_\delta$ we define the product $S_1 \cdot S_2$ for d = 2 if the
end point of S_1 equals the start point of S_2 resp., for d > 2,

if $S_1 \cap S_2$ is (d-2)-dimensional with opposite orientation.
For such S_i, i=1,2 the product $S_1 \cdot S_2$ is by definition
$S_1 \cup S_2$ as a set and its orientation is the one inherited
from those of S_1 and S_2.
For $S_1, \ldots, S_n \in H_\delta$ we give a recursive definition of $S_1, \ldots, S_n$
as $(S_1, \ldots, S_{n-1}) \cdot S_n$, whenever this is defined. We denote
by Σ_δ the set of all products of the form $S_1, \ldots, S_n$ with
$S_i \in H_\delta$.
For $S \in H_\delta$ we define S^{-1} as the element of H_δ equal to S as
a set, but with opposite orientation.

   Rem.: For d = 1, H_δ is simply identifiable with the set
of all points of M_δ. In the theory of lattice gauge fields,
d=2 the cells are called "plaquettes" and the faces are
called "links" or "bonds".

   A corsurface C on M_δ with values in a group G is a map
C from Σ_δ into G with
1) $C(S^{-1}) = (C(S))^{-1}$
2) $C(S_1 \cdot S_2) = C(S_1) C(S_2)$.
For d = 2 a cosurface is also called a "multiplicative curve
integrals". We use here the non commutative notation of
product. If G is abelian of course $C(S)^{-1}$ should be understood as $-C(S)$ and $C(S_1) \cdot C(S_2)$ as $C(S_1) + C(S_2)$.
Let $\Gamma_{\delta,G}$ be the set of all cosurfaces on M_δ. $\Gamma_{\delta,G}$ has a
natural measurable structure, whenever G has a measurable
structure, namely the one making all $S \to C(S)$, $S \in \Sigma_\delta$ measurable.

   A stochastic cosurface C on M_δ is a measurable map from
some probability space $(\Omega, \mathcal{A}, P)$ into $\Gamma_{\delta,G}$. For d = 1 a
stochastic cosurface is just a G-valued stochastic process
indexed by δZ, thus a stochastic cosurface is an extension
of the concept of stochastic process when the "time points"
$t \in \delta Z$ are replaced by d-1-dimensional hypersurfaces S of
$M_\delta = \delta Z^d$.

   We shall call complex any ordered n-tuple $K = \{S_1, \ldots, S_n\}$
$S_i \in \Sigma_\delta$, for some n. For d=1, K is just an n-tuple of

points. We shall use the notation $C(K) = \{C(S_1), \ldots, C(S_n)\}$ and look upon $C(K)$ as an element in G^K.

Two stochastic cosurfaces C, $\tilde{C}$ are called equivalent if the image probability measures under C, $\tilde{C}$ are equivalent as measures on G^K. We shall henceforth identify equivalent stochastic cosurfaces.

In the following we shall see that it is possible to define a probability measure on a suitable class of complexes. In fact the probability measure is first defined on regular complexes $K = \{S_1, \ldots, S_n\}$ which are <u>regular and saturated</u> in the sense that $S_i \cap S_j \subset \partial S_i \cap \partial S_j$ whenever $i \neq j$, $S_i, S_j \in K$ (regularity) and there exists a partition $D_K = \{A_1, \ldots, A_m\}$, for some $A_i \subset \mathbb{R}^d$ into connected and simply connected subsets A_i of $\mathbb{R}^d$ s.t. $\mathbb{R}^d$ is the union of the A_i and if $A_i \cap A_i \neq \emptyset$ for some $i \neq j$ then either $A_i \cap A_j$ is d-2-dimensional or else $A_i \cap A_j = \bigcup_k S_k^{i,j}$, for some $S_k^{i,j} \in \{S_1, \ldots, S_n\}$ and $\bigcup_{i \neq j} A_i \cap A_j = \bigcup_{i=1}^n S_i$. Assume for simplicity G is compact with countable base. The probability measure will be defined as a projective limit, starting from a definition on regular saturated complexes, which is given in terms of the Haar measure dx on G and a Markov convolution semigroup Q_t, $t \geq 0$ on G (semigroup of Markov kernels on G), with density $Q_t(x)$ s.t. $Q_t(x) \geq 0$, $\int Q_t(xy^{-1}) dy = 1$, $\int_G Q(xy^{-1}) Q_t(yz^{-1}) dy = Q_{s+t}(xz^{-1})$ and $Q_t(x) \to \delta_e(x)$ weakly as $t \downarrow 0$, with δ_e the Dirac measure at the origin $e \in G$. If G is non abelian (which is allowed for $d = 2$) we also assume Q_t invariant in the sense $Q_t(xy) = Q_t(yx) \; \forall x,y \in G$.

For any regular saturated complex $\tilde{K} = \{S_1, \ldots, S_n\}$ we define a probability measure $\mu_{\tilde{K}}^Q$ on $G^{\tilde{K}}$ by

$$d\mu_{\tilde{K}}^Q(C(\tilde{K})) \equiv Z^{-1} \prod_{A \in D_{\tilde{K}}} Q_{|A|} \left(\prod_{S \in \partial A \cap \tilde{K}} C(S) \right) dC(S_1) \ldots dC(S_n),$$

where ∂A is the topological boundary of A in $\mathbb{R}^d$, oriented with the normal of A pointing out of A. $|A|$ is the Lebesgue

volume of A and we set $Q_\infty = 1$. The product is an ordered
product along the oriented boundary ∂A of A. For any complex
$K = \{S_1,\ldots,S_n\}$ contained in a regular saturated complex
$\tilde{K} = \{\tilde{S}_1,\ldots,\tilde{S}_n\}$ (in the sense that all the elements in K can
be written as a product of the element of K or their inverses)
we define $d\mu_K^Q$ $(C(K) \equiv \int d\mu_K^Q$ $(C(\tilde{K})) \, \Pi \, dC(\tilde{S}_i)$, the integral
being on the $\tilde{S}_i$ not appearing in K and such that for
$S = \tilde{S}_1,\ldots, \tilde{S}_k$ we have $C(S) = C(\tilde{S}_1),\ldots,C(\tilde{S}_k)$.

μ_K^Q is then a measure on the set of all cosurfaces on K. The
semigroup property of Q together with the properties of
cosurfaces yield that (μ_K^Q, K) form a projective system on G^K,
if G is abelian in case $d > 2$. At least when G has a countable
base there is no problem then to get a probability space
$(\Omega, \mathcal{A}, P)$ and a stochastic cosurface C on it s.t. the image
of P under C(K) is μ_K^Q. $(\Omega, \mathcal{A}, P, C)$ is unique up to
equivalence. Moreover it is possible to show that C is a
Markov cosurface in the following natural sense.

For Λ a subset of M_δ, we shall denote by $\Sigma(\Lambda)$ the σ-algebra
generated by all stochastic variables C(S), $S \in \Sigma_\delta$, $S \subset \Lambda$.
We say that the stochastic cosurface C has the Markov
property i.e. is a Markov cosurface if for any regular
complex $K = \{S_1,\ldots,S_m\}$ with the properties that
$S_{i_1} \cup \ldots \cup S_{i_k}$, $S_{i_k} \in K$, $n=1,\ldots,\ell$ splits $\mathbb{R}^d - K$ into 2
connected components $M_K^\pm$ s.t. some of the S_i are in M_K^+ and
some are in M_K^- and $E(f_+ f_- | \Sigma(S)) = E(f^+ | \Sigma(S)) E(f^- | \Sigma(S))$ for
any $f^\pm$ which are $\Sigma(M_K^\pm \cup S)$-measurable and bounded. E.g. for
$d=2$, $m=3$, $\ell=1$, $i_1 = 2$, $S_2 = \{(x^1,x^2) \in \mathbb{R}^2 | x^2 = 0\}$ this is
conditional independence of the curve variables $C(S_i)$,
$i = 1$ resp. 3 in $x^2 > 0$ resp. $x^2 < 0$, given C(S). That
$(\Omega, \mathcal{A}, P, C)$ has this (global) Markov property is a simple
verification from the definition. Hence we have the following

Theorem 1   Let G be a compact group with countable base,
abelian for $d > 2$. Then μ_K^Q as defined above is a projective

family of probability measures. There exists a probability space $(\Omega, \mathcal{A}, P)$ and a stochastic cosurface C s.t. μ_K^Q are the finite dimensional distributions (marginals) of P. $(\Omega, \mathcal{A}, P, C)$ is unique up to equivalence. C has the global Markov property.

If we wish to stress the dependence of the Markov cosurface $(\Omega, \mathcal{A}, P, C)$ on Q, the volume measure $|A|$ on $\mathbb{R}^d$ and the orientation σ on $\mathbb{R}^d$ we write $C^{Q, \lambda, \sigma}$ for C. Using the invariance of $|A|$ under isometries of $\mathbb{R}^d$ (i.e. piecewise smooth transformations of $\mathbb{R}^d$ leaving λ and σ invariant) and the definition of μ_K^Q we have easily that $C^{Q, \lambda, \sigma}$ is invariant under isometries of $\mathbb{R}^d$ which carry M_δ into itself. Moreover $C^{Q, \lambda, \sigma} = C^{\tilde{Q}, \lambda, -\sigma}$ with $\tilde{Q}_t(x) \equiv Q_t(x^{-1})$; in particular if $\tilde{Q}_t = Q_t$, i.e. Q_t is reflection symmetric, then C is independent of σ.

Remark. Under a certain "joint measurability condition" (in ω and S) for a stochastic Markov cosurface $C(S, \omega)$ invariant under isometries of $\mathbb{R}^d$ one can prove a converse of the above result, see [67].

It is not difficult to extend all the preceding results to the case where M_δ is replaced by an oriented Riemannian manifold of dimension d. H_δ is then the family H_M of all oriented piecewise smooth connected $(d-1)$-dimensional hypersurfaces in M which are closed sets and do not have self-intersections. Σ_M is then defined accordingly as Σ_δ, with H_δ replaced by H_M. The set $\Gamma_{M, G}$ of all cosurfaces on M replaces then $\Gamma_{\delta, G}$. All other definitions hold with M_δ replaced by M (in the definition of regular saturated complex we also replace, of course, $\mathbb{R}^d$ by M). In the definition of μ_K^Q and μ_K^Q we replace the Lebesgue volume $|A|$ by $\lambda(A)$, where λ is the Riemannian volume on M. In the discussion about the dependence of $C^{Q, \lambda, \sigma}$ on λ, σ the isometries should be now understood as piecewise smooth transformations of M leaving λ and σ invariant.

The Theorem and the Remark following it hold then for this
situation, yielding a non commutative (G-valued) analogue
of a multidimensional time homogeneous Markov field (which
can be looked upon, as described in Sect. II, as a symmetric
Markov process, given by a Dirichlet form, on an infinite
dimensional space). We shall call the object of the Theorem
in this case a Markov cosurface over M, with values in G.

We shall remark that it is possible to look upon the
Markov cosurface $(\Omega, \mathcal{A}, P, C)$ over $M = \mathbb{R}^d$ as a limit of
Markov cosurfaces over a lattice $M_\delta = \delta Z^d$ as $\delta \downarrow 0$.

In fact, let Q_t be a convolution semigroup of Markov kernels
on G (compact, and abelian in case d > 2).
Let $K = \{S_1, \ldots, S_m\}$, $S_i \in \Sigma_\delta$ be any complex over M . Define
μ_K^Q using $Q = Q_t$, starting for (1), as above, with $Q_{|A|}$
replaced by $Q_{\beta|A|}$, with $|A|$, as before, the Riemann-
Lebesgue volume of A.
Using the Kolmogorov construction leading to Theor. 1 we get
a probability measure P_δ as projective limit of the μ_K^Q (we write
δ to underline the dependence on δ). Let $\delta = \delta(n) = \delta_0/2^n$,
for some fixed $\delta_0 > 0$ and with $n \in \mathbb{N}$. We can consider
$P_{\delta(n)}$ as a measure on $(\Omega_M, \mathcal{A}_M)$, the underlying measurable
space to the measure P_Q of the remark following Theor. 1,
by using the natural embeddings of M_δ and Σ_δ into $M = \mathbb{R}^d$
resp. Σ_M, and we then get that $P_{\delta(n)}$ converges weakly as
$n \to \infty$ if β is chosen to be n dependent, namely $\beta = 2^{nd}$ (the
limit probability measure is then the one $(\Omega_Q, \mathcal{A}_Q, P_Q)$
obtained from Theor. 1 with $Q = Q_t$).

For d = 2 there is an interesting connection between
the measures P_δ resp. P_Q and gauge fields on the lattice
M_δ resp. on $\mathbb{R}^2$. To describe this connection, let us introduce,
for general d, the probability measure on Markov cosurfaces
C(S) attached to complexes $K = \{S_1, \ldots, S_n\}$, S_i cells of M_δ,
by

$$d\mu_{\beta,U}^\delta(C(K)) = Z^{-1} \exp\left[-\beta \sum_{i=1}^{n} U(C(S_i))\right] dC(S_1)\ldots dC(S_n),$$

where Z is the normalizing factor and $\beta > 0$, U being an invariant $(U(xy) = U(yx), \forall x,y \in G)$ real-valued function on G.

There is a projective limit measure μ_δ such that the marginals of μ_δ are given by $\mu_{\beta,U}^Q$.

For $d = 2$, G a Lie group, μ_δ is the Gibbs measure describing lattice gauge fields on the lattice M_δ.

In [67], [68] it is shown, that at least when $G = U(1)$ or Z_2, (for arbitrary d), or $G = SU(2)$, for $d = 2$ and U is chosen as the real part of a character on G, there is a suitable choice $\beta(n)$ of β, depending on n, such that

$$\int f(C(F_1^\delta),\ldots, C(F_K^\delta)) \, d\mu_{\beta(n),U}^{\delta/2^n}$$

converges as $n \to \infty$ for any real-valued bounded measurable function f on G^{2dk}, with F_i^δ the faces of a finite set of cells $\gamma_1,\ldots,\gamma_k$ of M_δ, $\delta \overset{>}{=} 0$ fixed. The limit is then

$\int f dP_Q$, with P_Q as in the Remark following Theor. 1, with Q_t being a Markov semigroup of density kernels on G, constructed in a suitable way from U and β. E.g. for $d = 2$, $G = SU(2)$, $U(g) = \sin t/\sin t/2$, $t \in [0,2\pi)$,

$e^{\pm it/2}$ the eigenvalues of $g \in SU(2)$, $\beta(n) = 2^{2n}$ we have

$Q_t = e^{\frac{t}{2}\Delta}$, with Δ the Laplace-Beltrami operator on G.

Since for $d = 2$ the measure μ_δ is the Gibbs state of gauge fields and the above limit $n \to \infty$ describes the operation of "going to the continuum", we see that the measure P_Q describes the "continuum limit of gauge fields". Thus we also see that our Markov cosurfaces can be obtained as limits of cosurfaces attached to lattice gauge fields models, for $d = 2$, and for arbitrary $d > 2$ as limit of cosurfaces attached to certain lattice fields μ_δ.

Remark: In the case $M = \mathbb{R}^d$ one can associate to the covariant Markov cosurfaces $(\Omega_Q, \mathcal{A}_Q, P_Q; C)$ a Markov semigroup e^{-tH}, $t > 0$ acting on a Hilbert space of functions. H can be interpreted as the Hamiltonian of a relativistic theory, in which the relativistic quantum fields are associated with

d-1-dimensional hypersurfaces, rather than points. This
theory satisfes all postulates ([69]) for a relativistic
theory, see [67], [68] for details.

        Let us finally consider the case where G is a connected
compact Lie group. In this case there is a nice relation of
a Markov cosurface $(\Omega_Q, \mathscr{A}_Q, P_Q, C)$ with solutions of stochastic
equations of the form da = ξ, where a is a d-1-form on M
with values in the Lie algebra g of G and ξ is an infinitely
divisible generalized random field on M with values in g
This connection is best described in the abelian case G = T^ν
(the ν-dimensional torus), but holds also for G non abelian,
see [67].
Let thus G = T^ν and let C be a stochastic cosurface on a
Riemannian manifold M with values on G. Under a general
assumption of "joint measurability" it is possible to write
$C(S)$, $S \in \sum$ in the form

$$C(S) = e^{2\pi i \int_S a}, \quad \text{with } a = (a^1, \ldots, a^\nu)$$

a $\mathbb{R}$-valued d-1-form on M ("d-1-vector form"), where $e^{2\pi i \int_S a}$

stands for $(e^{2\pi i \int_S a^1}, \ldots, e^{2\pi i \int_S a^\nu})$ and the integrals are
understood as linear functionals, as currents, in the Rahm's
sense). The following theorem then holds

        Theorem 2. a) If C is a T^ν-valued, jointly measurable
Markov cosurface on M invariant under isometries of M, then
$C(S) = e^{2\pi i \int_S a}$, with a solution of the stochastic differential
equation da = ξdx, with ξ an infinitely divisible generalized
random field on M with values in $\mathbb{R}^\nu$ and dx the Rieman-
Lebesgue volume on M.
b) Conversely, given such a ξ, the stochastic equation
da = ξdx has a solution a such that $e^{2\pi i \int_S a}$ is a T^ν-valued
invariant Markov cosurface.

        The proof of a) uses the fact of the Remark following
Theor. 1, by which such a C has distribution $(\Omega_Q, \mathscr{A}_Q, P_Q)$ for
some Q. In the case $Q_t = e^{\frac{t}{2}\Delta}$, e.g., we have that $\int_S a^j$ for
S = A is distributed as $\int_0^{|A|} X_\tau d\tau$ with X_τ white noise on $\mathbb{R}$.

Thus, for any $\beta \in \mathbb{R}^\nu$, $E(e^{<\int_S a, \beta>}) = e^{-\frac{1}{2}|A|\beta^2}$, with $\beta^2 = <\beta, \beta>$, $<,>$ being the scalar product in $\mathbb{R}^\nu$. But this identifies $< \int_A a, \beta>$ with $<\xi, \beta\chi_A>$, proving $da = \xi$.

Remark There is an extension of this result to G non abelian and to non invariant Markov cosurfaces [67]. Results on continuity properties of $S \to C(S)$ can also be obtained [67], [70].

Remark The connection, for $d = 2$, with gauge fields can be shortly put in the following form. For $S = \partial\gamma$, γ a cell, $C(S)$ is a quantized Wilson loop for a continuum gauge field theory. Let E be the principal fiber bundle with base space M and fiber G. Let a be the 1-form an M obtained from the Lie-algebra valued 1-form $\tilde{a}$ on E giving a connection on E. Then, for $G = T^\nu$, $C(S) = e^{2\pi i \int_S a}$ is the holonomy operator. $Da = da = F$ is the $G = T^\nu$-Yang-Mills fields strength curvature form, which is the above infinitely divisible generalized random field ξ.

In conclusion, in this section we have sketched an extension to the case of state space a group G of the concept of Markov random field on a manifold M. The extension has also provided the starting point for a stochastic calculus for $d - 1$ stochastic forms and for an extension of multiplicative stochastic integrals to multiplicative stochastic integrals of currents. Inasmuch as Markov random fields can be looked upon as Markov processes with infinite dimensional state space, our discussion in this section can be also looked upon as a non commutative extension of those made in the previous sections. In all cases we have seen a close interconnection of several problems and we might hope that progress in one domain will have implications also in the other ones.

Acknowledgements

The first author would like to thank Professor Ludwig Arnold
and Professor Peter Kotelenez for their kind invitation to
the workshop. Useful discussions with A. Kaufmann and M.
Koeck in Bochum are gratefully acknowledged, as well as
stimulating stays at the Universities of Bielefeld, Bochum,
Oslo and at the Centre for Interdisciplinary Research,
Bielefeld, the Bielefeld-Bochum Research Center Stochastic
Processes (BiBoS) and the Centre de Physique Théorique
CNRS, Marseille. The partial financial support of the Volkswagenstiftung and the Norwegian Research Concil for Science
and the Humanities is also gratefully acknowledged. We
thank Mrs. Mischke and Mrs. Richter for the skilfull typing.

S. Albeverio,   R. Høegh-Krohn,   H. Holden
Mathematisches Institut,   Matematisk Institutt,   Matematisk
Ruhr-Universität, Bochum   Universitetet i Oslo   Institutt
and   and   Universitetet
BiBoS-Research Center of   Université de Provence   i Oslo
the Volkswagenstiftung   et Centre de Physique
                             Théorique, Marseille

References

[1] M. Fukushima, Dirichlet forms and Markov processes,
     North Holland Kodansha (1980)

[2] M.L. Silverstein, Symmetric Markov processes, Lect.
     Notes Maths. 426, Springer, Berlin (1974)

[3] M.L. Silverstein, Boundary theory for symmetric Markov
     processes, Lect. Notes Maths. 516, Springer, Berlin (1976)

[4] M. Fukushima, Energy forms and diffusion processes, ZiF-
     Preprint, in preparation (to appear in Proc. Bielefeld,
     Project 2: Mathematics - Physics, Ed. L. Streit, World
     Publishing, Singapore)

[5] E. Nelson, Dynamical Theories of Brownian motion,
     Princeton University Press, 1970; Quantum Fluctuations,
     Princeton University, in preparation

[6] S. Albeverio, R. Høegh-Krohn, A remark on the connection
     between stochastic mechanics and the heat equation,
     J. Math. Phys. 15, 1745-1747 (1974)

[7] S. Albeverio, R. Høegh-Krohn, Quasi invariant measures, symmetric diffusion processes and quantum fields, pp. 11-59 in "Proceedings of the International Colloquium on Mathematical Methods of Quantum Fields Theory", Ed. CNRS (1976)

[8] S. Albeverio, R. Høegh-Krohn, L. Streit, Energy forms, Hamiltonians and distorted paths, J. Math. Phys. 18, 907-917 (1977)

[9] M. Fukushima, A decomposition of additive functionals of finite energy, Nagoya Math. J. 74, 137-168 (1979)

[10] M. Fukushima, On a representation of local martingale additive functionals of symmetric diffusions, pp. 110-118 in Stochastic Integrals, Ed. D. Williams, Lect. Notes Maths. 851, Springer, Berlin (1981)

[11] M. Röckner, N. Wielens, Dirichlet forms - closability and change of speed measure, ZiF-Preprint, to appear in "Infinite dimensional analysis and stochastic processes", Ed. S. Albeverio, Res. Notes Maths., Pitman

[12] K. Rullkötter, U. Spönemann, Dirichletformen und Diffussionsprozesse, Diplomarbeit, Bielefeld; and paper in preparation

[13] J. Brasche, Störungen von Schrödingeroperatoren durch Masse, Diplomarbeit, Bielefeld (1983); Perturbations of Schrödinger Hamiltonians by measures-self-adjointness and lower semiboundedness, ZiF-Preprint (1984) (to appear in J. Math. Phys.)

[14] W. Karwowski, J. Marion, On the closability of some positive definite symmetric differential forms on $C_0^\infty(\Omega)$, CNRS, Marseille, Preprint (1984)

[15] S. Albeverio, R. Høegh-Krohn, L. Streit, Regularization of processes and Hamiltonians, J. Math. Phys. 21, 1636-1642 (1980)

[16] S. Albeverio, S. Kusuoka, L. Streit, Convergence of Dirichlet forms and associated Schrödinger operators, BiBoS-Preprint (1984)

[17] M. Fukushima, On a stochastic calculus related to
     Dirichlet forms and distorted Brownian motions, Phys.
     Repts. 77, 255-262 (1981)

[18] I. G. Hooton, Dirichlet forms associated with hyper-
     contractive semigroups, Trans. Am. Math. Soc. 253,
     237-256 (1979)

[19] N. Wielens, Eindeutigkeit von Dirichletformen und we-
     sentliche Selbstadjungiertheit von Schrödingeroperatoren
     mit stark singulären Potentialen, Diplomarbeit, Biele-
     feld (1983)
     On the essential self-adjointness of generalized
     Schrödingeroperators, J. Funct. Anal., 1984

[20] L. Streit, Energy forms: Schrödinger theory, processes,
     Phys. Repts. 77, 363-375 (1981)

[21] S. Albeverio, R. Høegh-Krohn, Markov processes and
     Markov fields in quantum theory, group theory, hydro-
     dynamics and C^*-algebras, pp. 497-540 in "Stochastic
     Integrals", Ed. D. Williams, Lect. Notes Maths. 851,
     Springer, Berlin (1980)

[22] S. Albeverio, R. Høegh-Krohn, Some remarks on Dirichlet
     forms and their applications to quantum mechanics and
     statistical mechanics, pp. 120-132 in Functional Analy-
     sis in Markov-Processes, Ed. M. Fukushima, Lect. Notes
     Maths. 923, Springer, Berlin (1982)

[23] S. Albeverio, R. Høegh-Krohn, Diffusion fields, quan-
     tum fields and fields with values in groups in Adv.
     probability, Stochastic Analysis, Ed. M. Pinsky,
     Dekker, New York (1984)

[24] R. Carmona, Processus de diffusion gouvernés par la
     forme de Dirichlet de l'opérateur de Schroedinger,
     Sém. Prob. XIII, Lect. Notes Maths. 721, Springer,
     Berlin (1979)

[25] S. Albeverio, F. Gesztesy, W. Karwowski, L. Streit, On
     the connection between Schrödinger- and Dirichlet
     forms, BiBoS-Preprint (1984)

[26] M. Nagasawa, Segregation of a population in an
     environment, J. Math. Biology 9, 213-235 (1980)

[27] S. Albeverio, M. Fukushima, W. Karwowski, L. Streit, Capacity and quantum mechanical tunneling, Comm. Math. Phys. 80, 301-342 (1981)

[28a] S. Albeverio, Ph. Blanchard, F. Gesztesy, L. Streit, Quantum mechanical low energy scattering in terms of diffusion processes, ZiF-Preprint 1984, to appear in Proc. II Enc. Math. Phys., France-Germany, Lect. Notes Maths., Springer (1984)

[28b] S. Albeverio, W. Karwowski, L. Streit, Some remarks on capacity, Green functions and Schrödinger operators, to appear in Ref. 11

[29] M. Fukushima, Markov processes and functional analysis, pp. 187-202 in Proc. Int. Math. Conf., L.H.Y. Chen et al, Edts., North Holland (1982)

[30] M. Fukushima, A note on irreducibility and ergodicity of symmetric Markov processes, pp. 200-208 in "Stochastic processes in quantum hteory and statistical physics", Edt. S. Albeverio et al, Lect. Notes Phys. 173, Springer, Berlin (1982)

[31] E. Carlen, Conservative diffusions and constructive approach to Nelson stochastic mechanics, Princeton Univ. Ph D Thesis (1984)

[32] W.A. Cheng, Sur la méchanique stochastique, RMA Strasbourg (1984)

[33] a) M. Fukushima, On distorted Brownian motions, pp. 255-262 in Ref. 20
     b) M. Fukushima, On absolute continuity of multidimensional symmetrizable diffusions, pp. 146-176 in Ref.22

[34] a) H. Nagai, Impulsive control of symmetric Markov processes and quasi-variational inequalities, Osaka J. Math. 20, 863-879 (1983)
     b) H. Nagai, Stopping problems of symmetric Markov processes and nonlinear variational inequalities, Preprint 1984

[35] M. Fukushima, A generalized stochastic calculus in homogeneization, pp.42-51 in "Quantum fields - Algebras, Processes", Proc. Symp. Bielefeld, 1978, Ed. L. Streit, Springer. Wien (1980)

[36] H. Föllmer, Dirichlet processes, pp. 476-478 in Ref. 21

[37] S. Kusuoka, Dirichlet forms and diffusion processes on
     Banach spaces, J. Fac. Sci. Univ. Tokyo 1A29, 79-95 (1982)

[38] M. Fukushima, M. Takeda, A transformation of a symmetric
     Markov processes and the Donsker-Varadhan theory, Osaka
     J. Math. 21, 311.326 (1984)

[39] M. Fukushima, M. Okada, On conformal martingales diffu-
     sions and pluripolar sets, J. Funct. Anal. 55, 377-388
     (1984)

[40] S. Albeverio, Ph. Blanchard, R. Høegh-Krohn, A stochastic
     model for the orbits of planets and satellites - The
     Titius-Bode law, Exp. Math. 4, 365-373 (1983)

[41] S. Albeverio, Ph. Blanchard, R. Høegh-Krohn, Diffusions
     sur une variété Riemannienne: barrières infanchissables
     et applications, Coll. L. Schwartz, Astérisque 1984

[42] S. Albeverio, Ph. Blanchard, Ph. Combe, R. Høegh-Krohn,
     R. Rodriguez, M. Sirugue, M. Sirugue-Collin, Magnetic
     bottles in a dirty environment. A stochastic model for
     radiation belts, ZiF-Preprint 1984

[43] S. Albeverio, R. Høegh-Krohn, The structure of diffusion
     processes, Marseille, Preprint

[44] S. Albeverio, R. Høegh-Krohn, Dirichlet forms and diffu-
     sion processes on rigged Hilbert spaces, Zeitschr. für
     Wahrscheinlichkeitstheorie verw. Gebiete 40, 1-57 (1977)

[45] S. Albeverio, R. Høegh-Krohn, Hunt processes and analytic
     potentials theory on rigged Hilbert spaces, Ann. Inst.
     H. Poincaré B 13, 269-291 (1977)

[46] a) S. Albeverio, R. Høegh-Krohn, Canonical quantum
        fields in two space-time dimensions, Preprint
     b) S. Albeverio, R. Høegh-Krohn, Topics in infinite
        dimensional analysis, pp. 278-302 in Lect. Notes
        Phys. 80, Springer (1978) (Ed. G.F. Dell'Antonio et al)

[47] Ph. Paclet, Espaces de Dirichlet et capacités fonctionnel-
     les sur des triplets de Hilbert-Schmidt, Sém. Paul Krée, 1977-1978
     No. 5, 1-36

[48] Ph. Paclet, Espaces de Dirichlet en dimension infinie, C.R. Ac.-Sciences Paris Ser. A 288, 981-993 (1979)

[49] M. Takeda, On the uniqueness of Markovian extension of diffusion operators on infinite dimensional spaces, ZiF-Preprint, 1984

[50] S. Albeverio, J.E. Fenstad, R. Høegh-Krohn, T. Lindstrøm. Non standard methods in stochastic analysis and mathematical physics, book in preparation (Acad. Press, (1985)

[51] B. Simon, The $P(\varphi)_2$ Euclidean (Quantum) field theory, Princeton Univ. (1974)

[52] S. Albeverio, R. Høegh-Krohn, Uniqueness and the global Markov property for Euclidean fields. The case of trigonometric interactions, Comm. Math. Phys. 68, 95-128 (1979)

[53] R. Gielerak, Verification of the global Markov property in some class of strongly coupled exponential interactions, J. Math. Phys. 24, 347-355 (1983)

[54] B. Zegarlinski, Uniqueness and the global Markov property for Euclidean fields: the case of general exponential interactions, Comm. Math. Phys., to appear

[55] S. Albeverio, R. Høegh-Krohn, Uniqueness and global Markov property for Euclidean fields and Lattice systems, pp. 303-329 in Quantum fields, Algebra Processes, Ed. L. Streit, Springer Verlag, Wien (1980)

[56] a) S. Albeverio, R. Høegh-Krohn, Stochastic methods in quantum field theory and hydrodynamics, Phys. Repts. 77, 193-214 (1981)
b) S. Albeverio, R. Høegh-Krohn, Local and global Markoff fields, Repts. Math. Phys. 13, 225-248 (1984)

[57] M. Röckner, Generalized Markov fields and axiomatic potential theory, Math. Ann. 264, 153-177 (1983)

[58] M. Röckner, Self-adjoint harmonic spaces and Dirichlet forms, Hiroshima Math. J., to appear

[59] M. Röckner, A Dirichlet problem for distributions and
     the construction of specifications for Gaussian
     generalized random fields, Mem. Am. Math. Soc., to appear

[60] M. Röckner, Generalized Markov fields and Dirichlet forms,
     Acta Applic. Math., to appear

[61] a) E.B. Dynkin, Markov processes and random fields, Bull
        Am. Math. Soc. 3, 975-999 (1980)
     b) T. Kolsrud, in preparation

[62] S. Albeverio, R. Høegh-Krohn, D. Testard, Factoriality
     of representations of the group of paths of SU(n),
     J. Funct. Anal. 57, 49-55 (1984)

[63] J. Marion, D. Testard, Energy representations of gauge
     groups coming from foliations, Luminy, Preprint (1984)

[64] S. Albeverio, R. Høegh-Krohn, Diffusion, quantum fields
     and groups of mappings, pp. 133-145 in Ref. 22

[65] L. Gross, Logarithmic Sobolov inequalities, Amer. J.
     Math. 97, 1061-1083 (1975)

[66] B. Lascar, Opérateurs pseudo-différentiels en dimension
     infinie. Etude de l'hypoellipticité et de la résolubilité
     dans des classes de fontions hölderiennes et de
     distribution pour des opérateurs pseudo-différentiels
     elliptiques, J. Anal. Math. 28, 39-104 (1978)

[67] S. Albeverio, R. Høegh-Krohn, H. Holden, in preparation

[68] S. Albeverio, R. Høegh-Krohn, H. Holden, Markov cosur-
     faces and gauge fields, Acta Phys. Austr. Suppl. 26,
     211-231 (1984)
     (Stochastic Methods and Computer Techniques in Quantum
     Dynmaics, Ed. H. Mitter, L. Pittner)

[69] E. Seiler, Gauge theories as a problem of construction
     quantum field theory and statistical mechanics, Lect.
     Notes Phys. 159, Springer, Berlin (1982)

[70] A. Kaufmann, in preparation

[71] J. Brasche, BiBoS-Preprint, in preparation

G. DA PRATO

MAXIMAL REGULARITY FOR STOCHASTIC CONVOLUTIONS AND
APPLICATIONS TO STOCHASTIC EVOLUTION EQUATIONS IN
HILBERT SPACES

1. INTRODUCTION

We are here concerned with the following problem:

$$(1.1) \quad dy = Ay\, dt + \sum_{i=1}^{N} B_i y\, dw_i(t) \quad , \quad y(0) = x$$

where A, B_1, B_2, ..., B_n are linear operators (generally unbounded) in a Hilbert space H and $w_1, w_2, ..., w_N$ are real
Brownian motions in a probability space (Ω, ϵ, P). When A
generates a strongly continuous semi-group in H and
$B_1, B_2, ..., B_N$ are bounded, problem (1.1) has been extensive-
ly studied. In this case Eq. (1.1) is written as an integral
equation

$$(1.2) \quad y(t) = e^{tA} x + \sum_{i=1}^{N} \int_0^t e^{(t-s)A} B_i y(s)\, dw_i(s)$$

which is solved by successive approximations (see [1], [2],
[5], [6], [7], [14], [15], [19], [20], [21], [22], [23],
[24], [28]).
When B_i are unbounded the most used approach is the variational one (see [29]).
If $N = 1$ and $B = B_1$ generate a group it is possible to
reduce Eq. (1.1) to a deterministic equation by setting
$v = e^{Bw_t} u$. In this case we obtain the problem:

L. Arnold and P. Kotelenez (eds.), Stochastic Space-Time Models and Limit Theorems, 41–52.
© 1985 by D. Reidel Publishing Company.

$$(1.3) \quad \begin{cases} v'(t,\omega) = e^{-Bw_t(\omega)}(A - \dfrac{B^2}{2})e^{Bw_t(\omega)}v(t,\omega) \\ \\ v(0,\omega) = x \qquad\qquad\qquad\qquad \forall \omega \in \Omega \end{cases}$$

and this problem can be studied via the abstract evolution equations theory. In this paper we will give some results using the regularizing properties of the Stochastic Convolution (see [8] and [9]).

In §2 we shall recall the main properties of Stochastic Convolution, in §3 we shall give some existence results.

2. STOCHASTIC CONVOLUTION

We assume here that A generates an analytic semi-group e^{tA} of negative type and that $f_i \in M^2(0,T;H)$, $i = 1,2,\ldots,N$. $M^2_w(0,T;H)$ represents the set of all stochastic processes adapted to $w(t) = w_1(t),\ldots,w_N(t))$ such that

$$E\int_0^T |g(s)|^2 \, dw_i(s) < +\infty \qquad i = 1,\ldots,N \quad .$$

We set

$$(2.1) \quad X_t = \sum_{i=1}^N \int_0^t e^{(t-s)A} f_i(s) \, dw_i(s) \quad .$$

X_t is called a stochastic convolution.

In the deterministic case the convolution $e^{tA} * f(f \in L^2(0,T;H))$ has the following regularizing property:

(2.2) $f \in L^2(0,T;H) \Rightarrow e^{tA} * f \in L^2(0;T,D(A))$

where $(e^{tA} * f)(t) = \int_0^t e^{(t-s)A} f(s)ds$.

Thus it is natural to expect that X_t takes values in some interpolation space between the domain $D(A)$ of A and H.

Let us recall the definition of interpolation spaces $D_A(\theta,2)$, $\theta \in]0,1[$ (see [4], [18], [27]). We say that x belongs to $D_A(\theta,2)$ if the following integral is finite:

(2.3) $|x|^2_{D_A(\theta,2)} = \int_0^\infty \xi^{1-2\theta} |A \, e^{\xi A} x|^2 \, d\xi$.

If A is self-adjoint negative then $D_A(\theta,2)$ is isomorphic to $D((-A)^\theta)$, moreover

(2.4) $|x|^2_{D_A(\theta,2)} = 2^{2\theta-2} \Gamma(2 - 2\theta)|(-A)^\theta x|^2_H$

We shall set in the sequel

(2.5) $|x|_{D((-A)^\theta)} = |(-A)^\theta x|$

Remark 2.1
   If $\theta \in]0,\frac{1}{2}[$ we have

(2.6) $|e^{tA}|_{\mathcal{L}(D_A(\theta,2))} \leq 1$.

Moreover, if $D(A) = D(A*)(A*$ being the adjoint of $A)$ there exists an operator A_1 and $\theta \in]0,\frac{1}{2}[$ such that $D_A = D_{A_1}(\theta,2)$; thus in this case there exists an equivalent scalar product in H with respect to which e^{tA} is a contraction

The following result is proved in [8] .

Proposition 2.2
Assume that $\theta \in]0,\frac{1}{2}[$, $f_i \in M_w^2(0,T;D_A(\theta,2))$, $i=1,2,\ldots,N$.
Then $X \in M_w^2(0,T;D_A(\theta + \frac{1}{2},2))$ and the following inequality
holds

$$(2.7) \quad \|X\|^2_{M_w^2(0,T;D_A(\theta+\frac{1}{2},2))} \leq \frac{1}{1-2\theta} \sum_{i=1}^{N} \|f_i\|^2_{M_w^2(0,T;D_A(\theta,2))} .$$

If A is self-adjoint we have the result

Proposition 2.3
Assume that A is self-adjoint negative, $\theta \in [0,\frac{1}{2}]$ and
$f_i \in M_w^2(0,T;D((-A)^\theta))$. Then $X \in M_w^2(0,T;D((-A)^{\theta+1/2}))$ and
the following inequality holds

$$(2.8) \quad \|X\|^2_{M_w^2(0,T;D((-A)^{\theta+1/2}))} \leq \frac{1}{2} \sum_{i=1}^{N} \|f_i\|^2_{M_w^2(0,T;D((-A)^\theta))}$$

Remark 2.4
Assume that the hypotheses of Proposition 2.1 hold.
Then since e^{tA} is a contraction in $D_A(\theta,2)$, $X(\cdot,\omega)$ is continuous in $D_A(\theta,2)$ a.e. in Ω (see [20] and [23]). For the
same reason if A is self-adjoint or if $D(A) = D(A*)$, $x(\cdot,\omega)$
is continuos in H a.e. in Ω.
In the general case we do not know if $X(\cdot,\omega)$ is continuous
in H. Moreover, if $f_i \in L_w^p([0,T];H)$ with $p > 2$, this is
true ([13]).

3.   EXISTENCE RESULTS

We are here concerned with the equation

$$(3.1) \quad y(t) = e^{tA} x + \sum_{i=1}^{N} \int_0^t e^{(t-s)A} B_i y(s) \, dw_i(s)$$

where e^{tA} is a strongly continuous semi-group in H.

Setting $z(t) = e^{-\omega t}y(t)$, $A_1 = A - \omega$, Eq. (3.1) reduces to

$$(3.2) \quad z(t) = e^{tA_1} x + \sum_{i=1}^{N} \int_0^t e^{(t-s)A_1} B_i z(s)\, dw_i(s) \quad ,$$

so it is not a restriction to assume e^{tA} of negative type.

Proposition 3.1

Let $\theta \in]0,\frac{1}{2}[$. Assume that:

$$(3.3) \begin{cases} \text{ a)} & e^{tA} \text{ is an analytic semi-group of negative type} \\ & \text{and } x \in D_A(\theta,2). \text{ } B_i \text{ maps } D_A(\theta + \frac{1}{2},2) \text{ into} \\ & D_A(\theta,2). \text{ Moreover} \\ \\ \text{ b)} & \exists \eta \in [0,1[\text{ such that} \\ \\ & \dfrac{1}{1-2\theta} \sum_{i=1}^{N} |B_i z|^2_{D_A(\theta,2)} \leq \eta |z|^2_{D_A(\theta+\frac{1}{2},2)} \quad \forall z \in D_A(\theta+\frac{1}{2},2) \end{cases}$$

then Eq. (3.1) has a unique solution $y \in M_w^2(0,T;D_A(\theta+\frac{1}{2},2))$.
Moreover $y(\cdot,\omega)$ is continuous in $D_A(\theta,2)$ for a.e. $\omega \in \Omega$.

Proof

We can write Eq. (3.1) in the form

$$y = e^{tA} x + \Gamma(y) \quad .$$

Let $Z = M_w^2(0,T;D_A(\theta + \frac{1}{2},2))$, then by Proposition 2.2 and
3.1 Γ maps Z into itself and $|\Gamma|_{\mathcal{L}(Z)} = \eta < 1$. Moreover
$e^{tA} x \in Z$, as can be easily checked, so that the conclusion
follows from the Contraction principle and Remark 2.3.

Remark that the solution y of Eq. (3.1) is not, in general, a strong solution of:

$$(3.4) \quad dy(t) = Ay(t)dt + \sum_{i=1}^{N} B_i y(t)\, dw_i(t) \quad , \quad y(0) = x$$

For this we need a stronger hypothesis on B_i. The proof of
the following proposition is similar to that of Proposition
3.1.

Proposition 3.2

Let $\theta \in]0,\frac{1}{2}[$. Assume that

(3.5) $\begin{cases} \text{a)} & e^{tA} \text{ is an analytic semi-group of negative type,} \\ & x \in D_A(\theta + \frac{1}{2},2) \\ \text{b)} & B_i \text{ maps } D_A(\theta + 1,2) \text{ into } D_A(\theta + \frac{1}{2},2) \text{ and } \\ & \exists \eta \in [0,1] \text{ such that} \\ & \frac{1}{1-2\theta} \sum_{i=1}^{N} |B_i z|^2_{D_A(\theta+\frac{1}{2},2)} \le \eta |z|^2_{D_A(\theta+1,2)} \quad \forall z \quad D_A(\theta+1,2) \end{cases}$

then Eq. (3.4) has a unique strong solution y with $y \in M^2_w(0,T;D_A(\theta + 1,2))$. Moreover $y(\cdot,\omega)$ is continuous in $D_A(\theta + \frac{1}{2},2)$ a.e. in Ω.

Let us consider now the case in which A is self-adjoint. Then, arguing as for the proofs of Proposition (3.1) and (3.2) we get the following:

Proposition 3.3

Let $\theta > 0$. Assume that

(3.6) $\begin{cases} \text{a)} & A \text{ is self-adjoint negative, } x \in D((-A)^{\theta}). \\ \text{b)} & B_i \text{ maps } D((-A)^{\theta + \frac{1}{2}}) \text{ into } D((-A)^{\theta}) \text{ and } \\ & \exists \eta \in [0,1[\text{ such that} \\ & \frac{1}{2} \sum_{i=1}^{N} |B_i z|^2_{D((-A)^{\theta})} \le \eta |z|^2_{D((-A)^{\theta+1/2})} \quad \forall z \in D((-A)^{\theta+\frac{1}{2}}) \end{cases}$

then Eq. (3.1) has a unique solution $y \in M^2_w(0,T;D((-A)^{\theta + \frac{1}{2}}))$. Moreover $y(\cdot,\omega)$ is continuous in $D_A(\theta,2)$ for a.e. $\omega \in \Omega$. Finally if $\theta \ge \frac{1}{2}$ then y is a strong solution of Eq. (3.4).

Remark 3.4

Hypothesis (3.3)-c is equivalent to $\gamma_\theta < 1$, where

(3.7) $\gamma_\theta = \frac{1}{1-2\theta} \operatorname{Sup}\{ \sum_{i=1}^{N} |B_i x|^2_{D_A(\theta,2)} ; |x|_{D_A(\theta+\frac{1}{2},2)} \le 1\}$

When A is self-adjoint it is not difficult to compute γ_θ whereas in the general case we can only give estimates for γ_θ (see Example 3). For this reason if $A = A_1 + C$, with A_1 self-adjoint and C continuous from $D_A(\theta + \frac{1}{2},2)$ into $D_A^1(\theta,2)$ it is more convenient to consider, instead to Eq. (3.1), the following one

$$(3.8) \quad y(t)=e^{tA_1}x + \int_0^t e^{(t-s)A_1}Cy(s)ds + \sum_{i=1}^N \int_0^t e^{(t-s)A_1}B_i y(s)dw_i(s)$$

that we can solve under hypotheses (3.6) (with A replaced by A_1).

We shall give now some examples, for sake of simplicity we shall only consider differential operators in one space variable. Many results can be easily generalized for operators in several variables.

Example 3.5 (Dirichlet problem)
    Let $H = L^2(0,\pi)$, $Au = u_{xx}$, $D(A) = H^2(0,\pi) \cap H_0^1(0,\pi)$, $Bu = bu_x$, with $b \in \mathbf{R}$. Equation (1.1) becomes

$$(3.9) \quad du = u_{xx} dt + bu_x dw_t , \quad u(0) = u_0$$

In this case A is self-adjoint, moreover we have (see for instance [26])

$$(3.10) \quad D((-A)^\theta) = \begin{cases} H^{2\theta}(0,\pi) & \text{if } \theta \in \left]0,\frac{1}{4}\right[\\ \{u \ H^{2\theta}(0,\pi); u(\cdot)=u(1)=0\} & \text{if } \theta \in \left]\frac{1}{4},1\right[\end{cases}$$

Moreover

$$(3.11) \quad |(-A)^\theta u|^2_{L^2(0,\pi)} = \frac{2}{\pi} \sum_{k=1}^\infty k^{4\theta} \left(\int_0^\pi u(x)\sin kx\, dx \right)^2 = |u|^2_{H^{2\theta}(0,\pi)}$$

Fix now $\theta \in [0,\frac{1}{4}[$ then $B_{\theta+1/2}$ maps $D((-A)^{\theta+\frac{1}{2}})$ into $D((-A)^\theta)$; moreover if $u \in D((-A)^{\theta+1/2})$

$$(3.12) \quad |u_x|^2_{D((-A)^\theta)} = \frac{2}{\pi} \sum_{k=1}^{\infty} k^{4\theta+2} (\int_0^\pi u(x) \cos kx \, dx)^2 =$$

$$= \frac{2}{\pi} \sum_{k=1}^{\infty} k^{4\theta+2} (\int_0^\pi u(x) \sin kx \, dx)^2 = |u|^2_{D((-A)^{\theta+1/2})}$$

it follows

$$(3.13) \quad \frac{1}{2}|Bu|^2_{D((-A)^\theta)} = \frac{b^2}{2} |u|^2_{D((-A)^{\theta+1/2})}$$

so that $\gamma_\theta = b^2/2$.

Thus if $|b| < \sqrt{2}$ and $u_0 \in H^{2\theta}(0,\pi)$, $\theta \in [0,\frac{1}{4}[$ there exists a unique solution $u \in M^2_w(0,T; H^{2\theta+1}(0,\pi) \cap H^1_0(0,\pi))$ to Eq. (3.1).

Remark that the condition $|b| < \sqrt{2}$ is the same that appears in the variational theory ([29]). This condition is quite natural, in fact if we consider the equation

$$(3.14) \quad du = u_{xx} dt + b u_x dw_t \quad , \quad u(0) = u_0$$

in $L^2(\mathbb{R})$ we can write down the following explicit solution

$$(3.15) \quad u(t,x) = e^{t(1-b^2/2)D_x^2} e^{w_t D_x} u_0(x)$$

Thus if $b < \sqrt{2}$ we have no solution and if $b = \sqrt{2}$ the solution does not belong to $H^{2\theta+1}$ (for general u_0).

Example 3.6   (Neumann Problem)

Let $H = L^2(0,\pi)$, $A_1 u = u_{xx}$, $D(A_1) = \{u \in H^2(0,\pi); u'(0) = u'(\pi) = 0\}$ $B = b u_x$, $b \in \mathbb{R}$. A_1 is self-adjoint and we have

$$(3.16) \quad D((-A_1)^\theta) = \begin{cases} H^{2\theta}(0,\pi) & \text{if } \theta \in]0,\frac{3}{4}[\\ \{u \in H^{2\theta}(0,\pi); u'(0)=u'(\pi)=0\} & \text{if } \theta \in]\frac{3}{4},1[\end{cases}$$

Fix now $\theta \in]\frac{1}{2}, \frac{3}{4}[$; then A_1 maps $D((-A_1)^{\theta+1/2})$ into $D((-A)_1^\theta)$; moreover proceeding as in Example (3.5) we can show that

$$(3.17) \quad \frac{1}{2}|Bu|^2_{D((-A_1)^\theta)} = \frac{b^2}{2}|u|^2_{D((-A_1)^{\theta+1/2})}$$

Thus if $|b| < \sqrt{2}$, $\theta \in]\frac{1}{2}, \frac{3}{4}[$ and $u_0 \in H^{2\theta}(0,\pi)$ there exists a unique strong solution $u \in M_w^2(0,T;H^{2\theta}(0,\pi))$ to the equation

$$(3.18) \quad du = u_{xx}\, dt + b\, u_x\, dw_t$$

(with boundary Neumann conditions)

<u>Example 3.7</u> (Non variational)
    Let $H = L^2(0,\pi)$, $A_2 u = du_{xx}$ with $\alpha \in C([0,\pi])$, $D(A_2) = H^2(0,\pi) \cap H_0^1(0,\pi)$. We assume

$$(3.19) \quad 0 < \epsilon \le \alpha \le k$$

Let $Bu = bu_x$, $b \in \mathbf{R}$, $D(B) = H_0^1(0,\pi)$. Remember that

$$(3.20) \quad D_{A_2}(\theta,2) = D_A(\theta,2)$$

where A is the operator defined in Example (3.5). Let us compute γ_θ; we have

$$(3.21) \quad |u|^2_{D(A_1)} \le k^2|u|^2_{D(A)} \le \frac{k^2}{2}|u|^2_{D(A_1)}$$

by interpolation we get

$$(3.22) \quad |u|_{D_{A_1}(\theta,2)} \le k^\theta|u|_{D_A(\theta,2)} \le (\frac{k}{\epsilon})^\theta|u|_{D_{A_1}(\theta,2)}$$

Using (3.12) we have

$$|Bu|^2_{D_{A_1}(\theta,2)} \le k^{2\theta}|Bu|^2_{D_A(\theta,2)} = k^{2\theta}2^{2\theta-2}\Gamma(2-2\theta)|Bu|^2_{D((-A)^\theta)}$$

$$\leq k^{2\theta} 2^{2\theta-2} \Gamma(2-2\theta) b^2 |u|^2_{D((-A)^{\theta+1/2})} = \frac{1-2\theta}{2} k^{2\theta} b^2 |u|^2_{D_A(\theta+\frac{1}{2},2)}$$

$$\leq \frac{1}{2} k^{2\theta} (1-2\theta)^{-2\theta-1} b^2 |u|^2_{D_{A_1}(\theta+\frac{1}{2},2)}$$

thus

$$(3.23) \quad \gamma_\theta \leq \frac{b^2}{2} \frac{k^{2\theta}}{\varepsilon^{2\theta+1}}$$

and if $\ |b| \ < \ 2 \dfrac{\varepsilon^{\theta+1/2}}{k^\theta} \ $ we can solve Eq. (3.1).

G. Da Prato

Scuola Normale Superiore

Pisa, Italy

REFERENCES

[1]   Arnold, L., Curtain, R., Kotelenez, P.: Nonlinear Sto-
       chastic Evolution Equations in Hilbert spaces, Report
       n° 17, Bremen University (1974).

[2]   Balakrishnan, A.: Stochastic Bilinear partial Differen-
       tial Equations, U.S. Italy Conference on Variable Struc-
       ture Systems, Oregon (1974).

[3]   Bensoussans, A., Temam, R.: Equations aux dérivées pat-
       tielles stochastiques non linéaires (1) Isr. J. Math.
       11 n° 1, 95-129 (1972).

[4]   Butzer, P., Berens, H.: Semigroup of Operators and Ap-
       proximation, Springer-Verlag (1967).

[5]   Chojnowska-Michalik, A.: Stochastic Differential Equa-
       tions in Hilbert Spaces and some of their applications,
       Thesis, Institute of Mathematics, Polish Academy of
       Sciences, 1977.

[6]   Curtain, R.F.: Stochastic Evolution Equations with ge-
        neral white noise disturbance. J. Math. Anal. Appl. 60
        (1977) 570–595.
[7]   Curtain, R.F., Pritchard, A.J.: Infinite Dimensional
        Linear Systems Theory, Springer–Verlag (1978).
[8]   Da Prato, G.: Regularity results of a convolution sto-
        chastic integral and applications to parabolic stocha-
        stic equations in a Hilbert space, Conference del Semi-
        nario di Matematica dell'Università di Bari (1982).
[9]   Da Prato, G.: Some results on linear stochastic evolu-
        tion equations in Hilbert spaces by the semi-groups
        methods, Stochastic Analysis and Applications 1, 1, 57–
        88 (1983).
[10]   Da Prato, G., Grisvard, P.: Equations d'évolution ab-
        straites non linéaires de type parabolique, Annali di
        Matematica pura e applicata (IV), vol. CXX p.P. 329–
        396 (1979).
[11]   Da Prato, G., Grisvard, P.: maximal regularity for evo-
        lution equations by interpolation and extrapolation.
        To appear in Jouranl of Functional Analysis.
[12]   Da Prato, G., Iannelli, M., Tubaro, L.: Some Results
        on Linear Stochastic Differential Equations in Hilbert
        Spaces, Stochastic, 6, 105–116 (1982).
[13]   Da Prato, G., Iannelli, M., Tubaro, L.: On the path regula-
        rity of a Stochastic Process in a Hilbert space defined
        by the Itô integral, Stochastic, 6, 315–322 (1982).
[14]   Da Prato, G., Iannelli, M., Tubaro, L.: Semi-linear
        stochastic differential equations in Hilbert spaces,
        Boll. U.M.I., 5, 168–185 (1979).
[15]   Dawson, D.A.: Stochastic evolution equations, Math.
        Biosciences, 15, 287–316 (1972).
[16]   Dunford, N., Schwartz, J.: Linear Operators II, Inter-
        science (1963).
[17]   Fleming, W.H.: Distribute parameter stochastic systems
        in population biology, Lec. Notes Econ. Math. Syst.,
        107, 179–191 (1975).
[18]   Grisvard, P.: Commutativité de deux foncteurs d'inter-
        polation et applications, J. Math. pures et appl., 45,

[19]   Ichikawa, A.: Linear stochastic evolution equations in
        Hilbert spaces, J. Diff. Equat., 28, 266-283 (1978).

[20]   Ichikawa, A.: Stability of semilinear evolution equa-
        tions. J. Math. Analysis and Appl., 90, 13-44 (1982).

[21]   Ichikawa, A.: Semilinear Stochastic Evolution Equations:

[22]   Kato, T.: Fractional powers of dissipative operators II,
        J. Math. Soc. Japan, 14 (1962).

[23]   Kotelenez, P.: A submartingale type inequality with ap-
        plications to stochastic evolution equations, Stochastics 8

[24]   Krylov,N.V., Rozovskii, B.L.: Itô equations in Banach
        spaces and strongly parabolic stochastic partial diffe-
        rential equations, Soviet Math. Dokl., 20, 1267-1271
        (1979).

[25]   Krylov, N.V., Rozovskii, B.L.: Stochastic evolution e-
        quations, Itogi Nauki i Tekhniki, Seyia Sovremennye
        Problemy Matenatiki, vol. 14, 71-146 (1979).

[26]   Lions, J.L.: Espaces d'interpolation et domaines de puis-
        sances fractionnaires d'opérateurs, J. Math. Soc. Japan,
        14, 234-241 (1962).

[27]   Lions, J.L, Magenes, E.: Problemes aux limites non-homo-
        genes et applications, Dunod, Paris, (1968).

[28]   Lions, J.L., Peetre, J.: Sur une classe d'espaces d'in-
        terpolation, Publ. I.H.E.S., 19, 5-68 (1964).

[29]   Metivier, M., Pistone, G.: Une formule d'isometrie pour
        l'integrale stochastique hilbertienne et équations d'é-
        volution linéaires stochastiques. Z. Wahrscheinlichkeit-
        stheorie verw. Gabiete, 33, 1-18 (1975).

[30]   Pardoux, E.: Equations aux dérivés partielles stochasti-
        ques non linéaires monotones, Thèse, Université, Paris
        XI (1975).

[31]   Pardoux, E.: Stochastic partial differential equations
        and filtering of diffusion processes, Stochastics, 3,
        (1979).

[32]   Yosida, K.: Functional Analysis and Semigroups, Springer-
        Verlag (1965).

Egbert Dettweiler

STOCHASTIC INTEGRATION OF BANACH SPACE VALUED FUNCTIONS

1. INTRODUCTION

Let M be a square integrable (real valued, right continuous) martingale relative to a certain filtration. Then the starting point of stochastic integration theory is the L^2-isometry between the space of square integrable predictable functions X (relative to the Doléans measure of M^2) and the space of the stochastic integrals $\int X dM$. This result extends without major difficulties to the case that M is Hilbert space valued and X belongs to a suitable space of operator-valued predictable functions.

More essential difficulties arise in the case that the state space of the stochastic integrals is a general Banach space. It was proved in [1] that an inequality of the form

$$\mathbb{E} \| \int X dM \|^2 \leq C \int \| X \|^2 d\mu \quad \text{(with a constant } C > o)$$

is valid in general if and only if the state space E of the integral process is 2-smoothable - in the sense that E has an equivalent 2-uniformly smooth norm (see e.g. [6]).

Since the class of 2-smoothable spaces is a rather restricted class of Banach spaces (to give an impression: only for $2 \leq p < \infty$ the L^p-spaces belong to this class), it seems to be an interesting problem to study the possibility of stochastic integration also on those Banach spaces having a bad geometry - in the sense that they are not 2-smoothable. This is the aim of the present paper.

The problems arising with Banach space valued stochastic integrals are mainly connected with the geometry of the state space of the integral process. The state space of the integrator process is not so important in this context. Indeed, even

53

L. Arnold and P. Kotelenez (eds.), Stochastic Space-Time Models and Limit Theorems, 53–79.
© 1985 by D. Reidel Publishing Company.

if the integrator process is one-dimensional, the main difficulties are the same.

For this reason we mostly restrict ourselves to the case that the integrator process is the one-dimensional Brownian motion. All essential problems remain with this restriction. Moreover, it turns out that generalizations to more general integrator processes (e.g. Banach space valued martingales, infinite-dimensional Wiener process etc.) can be derived from our results.

In the first part of the paper we introduce a notion of the stochastic integral of Banach space valued functions relative to Brownian motion which works on every (real, separable) Banach space. This notion has only some theoretical meaning, since on a general Banach space there are no nice criteria for stochastic integrability - as the classical isometry or an inequality of the form indicated above. Nevertheless, it shows that the stochastic integral of elementary functions can always be extended to a larger (Banach) space of functions.

The second part is concerned with the integrability of special classes of functions and applications to stochastic integral equations on arbitrary Banach spaces. We study especially the problem of giving conditions that a Lipschitz function $g: \mathbb{R}_+ \times E \longrightarrow F$ (E, F Banach spaces) extends to a Lipschitz function on the space of E-valued square integrable, predictable processes X in the sense that

(1.) the integral $\int g(s, X_s) d\xi_s$ always exists

(ξ := Brownian motion) and that

(2.) there exists a constant C>o such that for any two processes X,Y as above

$$\mathbb{E} \left\| \int (g(s,X_s) - g(s,Y_s)) d\xi_s \right\|^2 \leqslant C \, \mathbb{E} \int \|X_s - Y_s\|^2 ds.$$

Both properties of course hold for any classical Lipschitz function, if E and F are 2-smoothable, and it is well known from this case that property (2.) is intimately connected with the problem of proving existence and uniqueness of solutions of certain stochastic integral equations, where g is the integrand of the stochastic integral term.

In case that g does not depend on t there is a simple necessary and sufficient characterization that g has properties (1.) and (2.): g can be factorized in the form g = T∘f +x, where f is a classical Lipschitz function from E to a Banach space G and T is a special bounded linear operator from G to F which will be called 2-smooth. Similar but more complicated characterizations hold in the case that g depends on t.

It follows from these results that the class of 2-smooth operators (already mentioned in[6], see also[7]) is an interesting class from the viewpoint of infinite-dimensional stochastic integration, and we present some examples of operators belonging to this class.

Finally, we also use the class of 2-smooth operators to define for every Banach space E a reasonable large space of E-valued predictable processes X, for which the existence of the stochastic integral $\int X d\xi$ follows as in the classical situation by an inequality of the form

$$\mathbb{E} \ \| \int X \ d\xi \|^2 \leqslant C \ \mathbb{E} \int \| f(X) \|^2 ds \ .$$

2. GENERAL DEFINITION OF THE STOCHASTIC INTEGRAL

Let $(\Omega, \mathcal{F}, P)$ be a fixed probability space and let $(\mathcal{F}_t)_{t \geqslant 0}$ be a standard filtration of Ω , i.e. we assume that $\mathcal{F}_0$ contains all P-null sets of the P-completion of $\mathcal{F}_{\infty} := \mathfrak{6}(\mathcal{F}_t : t \geqslant 0)$ and that $(\mathcal{F}_t)$ is right continuous. Let $\xi = (\xi_t)$ denote a (one-dimensional) Brownian motion defined on Ω , and assume that $(\mathcal{F}_t)$ has the following additional properties: ξ is $(\mathcal{F}_t)$-adapted and $\xi_t - \xi_s$ is independent from $\mathcal{F}_r$ for $0 \leqslant r \leqslant s \leqslant t$.

Let $\mathcal{P}$ denote the $\mathfrak{6}$-algebra of predictable subsets of $\mathbb{R}_+ \times \Omega$. By definition $\mathcal{P}$ is generated by the family $\mathcal{R}$ of the predictable rectangles $]s,t] \times F \ (0 \leqslant s \leqslant t, F \in \mathcal{F}_s)$ or $\{0\} \times F \ (F \in \mathcal{F}_0)$.

Now let E be a real, separable Banach space. We call a function $f: \mathbb{R}_+ \times \Omega \longrightarrow E$ an elementary (predictable E-valued) function, if f is of the form $f = \sum_{k=0}^{n-1} x_k 1_{]t_k, t_{k+1}]} \times F_k$ ($x_k \in E, o \leq t_o < t_1 < \ldots < t_n$, $F_k \in \mathcal{F}_{t_k}$). By $\mathcal{E} := \mathcal{E}(E)$ we denote the vector space of all elementary functions. Let λ denote the Lebesgue measure on $\mathbb{R}_+$. If we identify functions in $\mathcal{E}$ which differ only on a $\lambda \otimes P$-null set, we can view $\mathcal{E}$ as a subspace of $L(E', L^2(\mathbb{R}_+ \times \Omega, \mathcal{P}, \lambda \otimes P))$ with its uniform operator norm, which we denote by $\| . \|_{(2)}$. Let $\mathbb{L}_{(2)}$ denote the completion of $\mathcal{E}$ with respect to $\| . \|_{(2)}$. Then the elements of $\mathbb{L}$ will be called (square integrable, predictable) <u>cylindrical func-tions</u>.

For $f \in \mathcal{E}$ of the above form the <u>stochastic integral</u> $\int f d\xi$ of f relative to ξ is defined as

$$\int f d\xi := \int f(t) d\xi_t := \sum_{k=0}^{n-1} x_k 1_{F_k} (\xi_{t_{k+1}} - \xi_{t_k}) .$$

Let $\| f \|_{2,\xi}$ denote the L^2-norm of $\int f d\xi$. Then the one-dimensional L^2-isometry implies $\| f \|_{2,\xi} \geq \| f \|_{(2)}$ and hence the completion $\mathbb{L}_\xi^2$ of $\mathcal{E}$ relative to the norm $\| . \|_{2,\xi}$ is contained in $\mathbb{L}_{(2)}$. The elements of $\mathbb{L}_\xi^2$ will be called ξ <u>-square integrable cylindri-cal functions</u> and for $f \in \mathbb{L}_\xi^2$ the stochastic inte-gral $\int f d\xi$ of f relative to ξ is defined as

$$\int f d\xi := \int f(t) d\xi_t := \lim_{n \to \infty} \int g_n d\xi ,$$

if (g_n) is a sequence in $\mathcal{E}$ with $\lim_{n \to \infty} \| f - g_n \|_{2,\xi} = o$.

2.1 <u>Proposition</u>. If $f \in \mathbb{L}_\xi^2$, then also $f 1_{[o,t]} \in \mathbb{L}_\xi^2$ for all $t \geq o$, and $\left(\int_o^t f(s) d\xi_s \right)_{t \geq o}$ (where of course $\int_o^t f(s) d\xi_s := \int 1_{[o,t]} f d\xi$) can be viewed (and we will do so in the following) as a <u>continuous</u> square integrable martingale - in the sense that there exists a continuous version.

<u>Proof</u>: Let (g_n) be a sequence in $\mathcal{E}$ with $\lim \| f - g_n \|_{2,\xi} = o$. For every $n \geq 1$ the process $\left(\int g_n 1_{[o,t]} d\xi \right)_{t \geq o}$ is a continuous martingale. This implies

$$\mathbb{E} \| \int 1_{[o,t]} (g_n - g_m) d\xi \|^2 \leq \| g_n - g_m \|_{2,\xi}^2$$

for all $n, m \in \mathbb{N}$ and hence $f 1_{[o,t]} \in \mathbb{L}_\xi^2$. The martingale property of $(\int f1_{[0,t]}d\xi)_{t \geqslant 0}$ follows from

$$\mathbb{E}[\int fd\xi \mid \mathcal{F}_t] = \lim_{n \to \infty} \mathbb{E}[\int g_n d\xi \mid \mathcal{F}_t] = \lim_{n \to \infty} \int g_n 1_{[0,t]} d\xi = \int_0^t fd\xi .$$

Now let $(\delta_k), (\varepsilon_k), (\eta_k)$ be sequences of positive numbers with the properties

$$\varepsilon_m = \sum_{k \geqslant m} \delta_k, \quad \lim_{m \to \infty} \varepsilon_m = 0 \text{ and } \sum_{k \geqslant 1} \delta_k^{-2} \eta_k^2 < \infty ,$$

and choose a sequence (n_k) in $\mathbb{N}$ such that $\|g_{n_{k+1}} - g_{n_k}\|_{2,\xi} \leqslant \eta_k$. Then we get

$$P\left[\sup_{t \geqslant 0} \|\int_0^t (g_{n_{k+1}} - g_{n_k})d\xi\| \geqslant \delta_k\right] \leqslant \delta_k^{-2}\eta_k^2$$

and hence by the Borel-Cantelli lemma

$$P\left[\exists m \; \forall l \geqslant k \geqslant m: \; \sup_{t \geqslant 0} \|\int_0^t (g_{n_l} - g_{n_k})d\xi\| \leqslant \varepsilon_k\right]$$

$$\geqslant P\left[\exists m \; \forall k \geqslant m: \; \sup_{t \geqslant 0} \|\int_0^t (g_{n_{k+1}} - g_{n_k})d\xi\| \leqslant \delta_k\right] = 1 .$$

This proves that there exists a continuous version for the martingale $(\int_0^t fd\xi)_{t \geqslant 0}$.

Now we extend the stochastic integral to spaces of cylindrical functions which are only locally integrable in the following sense:

A cylindrical function $f \in L(E', L^0(\mathbb{R}_+ \times \Omega, \mathcal{F}, \lambda \otimes P))$ is said to be <u>locally ξ-integrable</u>, if there exists an increasing sequence (T_n) of stopping times with $\lim_{n \to \infty} T_n = \infty$ a.s. such that

(i) $f1_{[0,T_n]} \in \mathbb{L}_\xi^2$ for all $n \geqslant 1$ and

(ii) $\int_0^t fd\xi := \lim_{n \to \infty} \int f1_{[0,t \wedge T_n]} d\xi$ exists a.s.
      for all $t \geqslant 0$.

$\mathbb{L}_\xi^{loc}$ will denote the space of all locally ξ-integrable cyl. functions. It is not difficult to see that there exists always a continuous version of $(\int_0^t fd\xi)_{t \geqslant 0}$ and in this sense $(\int_0^t fd\xi)_{t \geqslant 0}$ is a continuous local martingale.

2.2 <u>Proposition</u>. For every $f \in \mathbb{L}_\xi^{loc}$ there exist a sequence (g_n) in $\mathcal{E}$ and a double-sequence $(T_{n,k})$ of stopping times increasing in n and k and having the property $\lim_{k \to \infty} \sup_{n \geqslant 1} T_{n,k} = \infty$ a.s. such that

(1) for all $k \geqslant 1$, $(\int g_n 1_{[o,T_{n,k}]} d\xi)_{n \geqslant 1}$ is a Cauchy sequence in $\mathbb{L}_\xi^2$,

(2) for all $t \geqslant 0$, $(\int_o^t g_n d\xi)_{n \geqslant 1}$ is a Cauchy sequence in probability, and

(3) $f 1_{[o,T_k]} = \lim_{n \to \infty} g_n 1_{[o,T_{n,k}]}$ in $\mathbb{L}_{(2)}$, where $T_k = \sup_{n \geqslant 1} T_{n,k}$.

Conversely, suppose that (g_n) is a sequence in $\mathcal{E}$ and $(T_{n,k})$ is a double-sequence of stopping times as above such that (1) and (2) hold. Then (3) defines (!) an $f \in \mathbb{L}_\xi^{loc}$.

Proof: I. Let $f \in \mathbb{L}_\xi^{loc}$ and let (T_n) be a localizing sequence of stopping times for f such that conditions (i) and (ii) hold. Then for every fixed $n \geqslant 1$ there exists a sequence $(f_{n,k})_{k \geqslant 1}$ in $\mathcal{E}$ such that $\lim_{k \to \infty} f_{n,k} = f 1_{[o,T_n]}$ in $\mathbb{L}_\xi^2$. Now let (ε_n) be a given null-sequence of positive numbers and put $t_n = \varepsilon_n^{-1}$ for all $n \geqslant 1$. Then we can find for every $n \geqslant 1$ an m_n such that $P[T_{m_n} \leqslant t_n] \leqslant \varepsilon_n$ and then a k_n such that $\| f_{m_n,k_n} 1_{[o,T_{m_n}]} - f 1_{[o,T_{m_n}]} \|_{2,\xi}^2 \leqslant \varepsilon_n^3$. For every $t \leqslant t_n$ we get for the sequence (g_n) given by $g_n := f_{m_n,k_n}$:

$$\mathbb{E}\left[\inf\left(1, \sup_{0 \leqslant s \leqslant t} \| \int g_n 1_{[o,s]} d\xi - \int f 1_{[o,o]} d\xi \|\right)\right]$$

$$\leqslant P[T_{m_n} \leqslant t_n] + \int 1_{[T_{m_n} > t_n]} \inf\left(1, \sup_{s \leqslant t} \| \int_o^s g_n d\xi - \int_o^s f d\xi \|\right) dP$$

$$\leqslant \varepsilon_n + \mathbb{E}\left[\inf\left(1, \sup_{s \leqslant t} \| \int g_n 1_{[o,s \wedge T_{m_n}]} d\xi - \int f 1_{[o,s \wedge T_{m_n}]} d\xi \|\right)\right]$$

$$\leqslant 2\varepsilon_n + P\left[\sup_{s \leqslant t} \| \int g_n 1_{[o,s \wedge T_{m_n}]} d\xi - \int f 1_{[o,s \wedge T_{m_n}]} d\xi \| \geqslant \varepsilon_n\right]$$

$$\leqslant 2\varepsilon_n + \varepsilon_n^{-2} \mathbb{E}\| \int g_n 1_{[o,T_{m_n}]} d\xi - \int f 1_{[o,T_{m_n}]} d\xi \|^2 \leqslant 3\varepsilon_n ,$$

where the last line follows from Doob's inequality. This proves that $(\int_o^t g_n d\xi)_{n \geqslant 1}$ is a Cauchy sequence relative to the Fréchet norm $\| . \|_o$ on $L^o(\Omega, \mathcal{F}, P; E)$ defined by $\| X \|_o := \mathbb{E} \inf(1, \|X\|)$, i.e. we have proved that (2) holds. To prove (1) and (3) we define for any $g \in \mathbb{L}_\xi^{loc}$ and every $c \geqslant 0$ the following stopping times:

$$T_c(g) := \inf\{t \geqslant 0 : \|\int_0^t g\,d\xi\| > c\} \quad \text{and}$$
$$T_{c-}(g) := \inf\{t \geqslant 0 : \|\int_0^t g\,d\xi\| \geqslant c\}.$$

Then it follows from the continuity of the process $(\int f 1_{[0,t]}\,d\xi)_{t \geqslant 0}$ that for every $c \geqslant 0$

$$(*) \quad T_{c-}(f) \leqslant \varliminf T_{c-}(g_n) \leqslant \varlimsup T_c(g_n) \leqslant T_c(f).$$

We put $T_{n,k} := T_{k-}(f) \wedge \inf_{m \geqslant n} T_k(g_m)$ for every $n \geqslant 1, k \geqslant 1$. Then for every fixed k the sequence $(\int_0^{T_{n,k}} g_n\,d\xi)_{n \geqslant 1}$ is uniformly bouded by k and it follows from $(*)$ and (2) that (1) and (3) hold.

II. If (g_n) is a sequence in $\mathcal{E}$ such that for a double-sequence $(T_{n,k})$ as above conditions (1) and (2) hold, then for every $k \geqslant 1$ we put

$$f_k := \lim_{n \to \infty} g_n 1_{[0,T_{n,k}]} \quad (\text{in } \mathbb{L}_{(2)}).$$

It follows from (1) that

$$f_j 1_{[0,T_k]} = f_k \quad \text{for } k < j \text{ (where } T_k := \sup_{n \geqslant 1} T_{n,k} \text{)},$$

hence the cylindrical function f defined by f_k on $[0,T_k]$ is well defined, and it is not difficult to see that f has properties (i) and (ii), i.e. f is locally ξ-integrable.

Remark: $\mathbb{L}_\xi^{loc}$ can also be obtained (as $\mathbb{L}_\xi^2$) as a completion of $\mathcal{E}$ relative to a suitable Fréchet norm. For every $g \in \mathcal{E}$ we put

$$\|g\|_{0,\xi} := \sup_{t \geqslant 0} t^{-1} \mathbb{E}[\inf(1, \sup_{0 \leqslant s \leqslant t} \|\int_0^s g\,d\xi\|)].$$

Then $\mathbb{L}_\xi^{loc} = (\mathcal{E}, \|.\|_{0,\xi})^{\sim}$.

Propositions 2.1 and 2.2 show that it is possible to define on an arbitrary Banach space E the stochastic integral for a space of (cylindrical) functions which is always larger than $\mathcal{E}$ (since $\mathbb{L}_\xi^2$ is a Banach space). There are two problems hidden behind our definition of the stochastic integral. The minor problem is that $\mathbb{L}_\xi^2$ need not be a space of E-valued functions. This could be avoided for example in the following way. Consider instead of $\mathbb{L}_\xi^2$ the smaller Banach space

$$\mathbb{L}_\xi^{(2)} := \mathbb{L}_\xi^2 \cap L^2(\mathbb{R}_+ \times \Omega, \mathcal{P}, \lambda \otimes P; E)$$

(with the norm $\| \cdot \|_{(2),\xi} := \max(\| \cdot \|_{2,\xi}, \| \cdot \|_2)$) which is again a Banach space containing ξ .

The major problem consists in the lack of a general criterion to decide whether a given function belongs to $\mathbb{L}_\xi^2$, i.e. can be approximated by a sequence of elementary functions in the $\mathbb{L}_\xi^2$-norm. In the case of a Hilbert space this criterion is of course given by the isometry between $\mathbb{L}_\xi^2$ and $L^2(\lambda \otimes P, E)$. If E is not a Hilbert space, then this isometry is no longer valid. More precisely, we have the following result (which is proved in [1]):

2.3 <u>Theorem.</u>

(1) For a Banach space E the inclusion $\mathbb{L}_\xi^2 \supset L^2(\lambda \otimes P, E)$ holds if and only if E is 2-smoothable (i.e. there exists an equivalent norm $| \cdot |$ on E such that $(E, | \cdot |)$ is 2-uniformly smooth, see [6] or [7]).
(2) The inclusion $\mathbb{L}_\xi^2 \subset L^2(\lambda \otimes P, E)$ holds if and only if E is 2-convexifiable (i.e. there exists an equivalent norm $| \cdot |$ on E such that $(E, | \cdot |)$ is 2-uniformly convex).
(3) $\mathbb{L}_\xi^2 = L^2(\lambda \otimes P, E)$ if and only if E is isomorphic to a Hilbert space.

The theorem indicates that for not 2-smoothable spaces it will be a problem to give general criteria that given functions belong to $\mathbb{L}_\xi^2$. This will be the subject of the following paragraphs.

3.   LIPSCHITZ FUNCTIONS ON BANACH SPACES

Let F be a second Banach space. Then $\text{Lip}(\mathbb{R}_+ \times E, F)$ will denote the space of all Borel measurable functions $g: \mathbb{R}_+ \times E \longrightarrow F$ with the property

(Lip) $\sup_{t \geq 0} \| g(t,o) \| < \infty$ and there exists a constant

$c \geq o$ such that for all $x, y \in E$

$\sup_{t \geq 0} \| g(t,x) - g(t,y) \| \leq c \| x - y \|$.

The subspace of Lipschitz functions which do not depend on t will be denoted by $\text{Lip}(E, F)$. For every $g \in \text{Lip}(\mathbb{R}_+ \times E, F)$ let $L(g)$ be the smallest constant $c \geq \sup \| g(t,o) \|$ such that (Lip) holds and suppose for a moment that F is 2-smoothable. Then it is known (see [1]) that there exists a constant $C \geq o$

such that for all $g \in \text{Lip}(\mathbb{R}_+ \times E, F)$ and all $X, Y \in L^2(\lambda \otimes P, E)$

$$\mathbb{E} \| \int g(t, X_t) d\xi_t - \int g(t, Y_t) d\xi_t \|^2$$

$$\leqslant c^2 \, \mathbb{E} \int \| g(t, X_t) - g(t, Y_t) \|^2 dt$$

$$\leqslant c^2 L(g)^2 \, \mathbb{E} \int \| X_t - Y_t \|^2 \, dt \; .$$

In the notation of section 2 this means

$(*)$ $\| g(.,X.) - g(.,Y.) \|_{2,\xi} \leqslant CL(g) \| X - Y \|_2$,

i.e. $g \in \text{Lip}(L^2(\lambda \otimes P, E), \mathbb{L}^2_\xi(F))$.

It is well known that inequality $(*)$ is the essential inequality in proving existence and uniqueness for the solutions of diffusion equations, where g is the function occuring in the diffusion term (see[3],[4],[5]and in the Banach space case e.g.[2]). If F is not necessarily 2-smoothable, then in general $(*)$ is not valid because of the following two problems:

Problem 1: If $g \in \text{Lip}(\mathbb{R}_+ \times E, F)$ and $X \in L^2(\lambda \otimes P, E)$, then not necessarily $g(X) := g(.,X.) \in \mathbb{L}^2_\xi$, i.e. the stochastic integral $\int g(X) d\xi$ need not exist.

Problem 2: Even if $g(X) \in \mathbb{L}^2_\xi$, then inequality $(*)$ need not be valid.

For these reasons we will now study a reasonable large subclass of Lipschitz functions, for which both problems have a positive answer.

We introduce the following notational conventions. Let π denote the family of all finite sequences $\Delta = (t_k)_{0 \leqslant k \leqslant n}$ in $\mathbb{R}_+$ with $t_k < t_{k+1}$ for $0 \leqslant k \leqslant n-1$. π_t will denote the subfamily of all $\Delta \in \pi$ with $t_0 \geqslant t$. For every $\Delta \in \pi$ we put

$$\varphi^\Delta := \sum_{k=0}^{n-1} t_k 1_{]t_k, t_{k+1}]} \; ,$$

and for any function f defined on $\mathbb{R}_+$ we denote by f^Δ the function $f \circ \varphi^\Delta$.

Let us call a function $f: \mathbb{R}_+ \longrightarrow E$ ξ-continuous, if the following two conditions hold:

(i) $f \in \mathbb{L}^2_\xi(E)$ and $\lim\limits_{|\Delta| \to 0} \| f^\Delta - f \|_{2,\xi} = 0$ (where as
      usual $|\Delta| := \max(t_{k+1}-t_k)$ for $\Delta = (t_k)_{0 \le k \le n}$).

(ii) $\lim\limits_{h \downarrow 0} h^{-1} \mathbb{E} \| \int_t^{t+h} (f(s)-f(t)) \, d\xi_s \|^2 = 0$ for all $t \ge 0$.

By $C_\xi(\mathbb{R}_+, E)$ we will denote the space of all ξ-
continuous E-valued functions on $\mathbb{R}_+$.

   Now we will say that a Borel function g: $\mathbb{R}_+ \times E$
$\longrightarrow$ F is a ξ-<u>Lipschitz function</u>, if g has the
following properties:

(1) $g(.,x) \in C_\xi(\mathbb{R}_+, F)$ for all $x \in E$, and

(2) there exists a constant $C \ge 0$ such that for
    all $X,Y \in \check{\mathbb{C}}(E)$ and all $Z \in \xi(\mathbb{R})$:

$$\mathbb{E} \| \int (g(X)-g(Y)) \, Z \, d\xi \|^2 \le C^2 \, \mathbb{E} \int (Z \| X - Y \|)^2 d\lambda.$$

We denote by C(g) the smallest constant C for which
inequality (2) holds, and we will write $\mathrm{Lip}_\xi(\mathbb{R}_+ \times E, F)$
for the space of all ξ-Lipschitz functions. Finally,
$\mathrm{Lip}_\xi(E,F)$ will denote the subspace of all ξ-Lip-
schitz functions, which do not depend on t (so that
condition (1) is trivially fulfilled).
   We will now look for simpler characterizations
of the space $\mathrm{Lip}_\xi(\mathbb{R}_+ \times E, F)$. First we prove a renorm-
ing lemma, which will be essential for the later
characterizations (cf.[6], theorem 3.1 for the
origin of the main trick in the proof).

3.2 <u>Lemma</u>. For every $g \in \mathrm{Lip}_\xi(\mathbb{R}_+ \times E, F)$ there exists
a family $(n_t)_{t \ge 0}$ of norms on F with the following
properties:

(i) $\| x \| \le n_t(x) \le n_s(x) \le \sqrt{2} \| x \|$ for all $x \in F$ and
      $0 \le s \le t$, and

      $\lim\limits_{s \downarrow t} n_s(x) = n_t(x)$ for all $x \in F$, $t \ge 0$.

(ii) There exists a constant $D = D(g)$ such that for
      a given $N(0,1)$-distributed Gaussian random
      variable ξ, for all $t \ge 0$, $x \in F$ and $y, z \in E$

$$\mathbb{E} \, n_t(x + (g(t,y)-g(t,z))\xi)^2 \le n_t(x)^2 + D^2 \| y-z \|^2.$$

<u>Proof</u>: We put $\mathfrak{F}_{s,t} := \mathfrak{S}(\xi_r - \xi_s; \; s \leqslant r \leqslant t)$ and denote by $\mathcal{P}_s$ the σ-algebra of predictable subsets of $[s, \infty] \times \Omega$ defined by the filtration $(\mathfrak{F}_{s,t})_{t \geqslant s}$. We will use the notation $X \in \mathcal{E}_s(E)$ to express that $X \in \mathcal{E}(E)$ is $\mathcal{P}_s$-predictable.

Now we define for every $t \geqslant o$, $x \in F$

$$n_t(x)^2 := \sup_{X,Y,Z} \; \mathbb{E}\|x + \int_t^\infty Z(g(X)-g(Y))d\xi\|^2 - 2C(g)^2 \int_t^\infty \mathbb{E}(Z\|X-Y\|)^2 d\lambda \; ,$$

where the supremum is taken over all $X,Y \in \mathcal{E}_t(E)$ and $Z \in \mathcal{E}_t(\mathbb{R})$. We will prove that the family (n_t) has properties (i) and (ii).

Let us first show that property (ii) holds. We put $G := g(\cdot,y)-g(\cdot,z)$ and prove that for every fixed $\varepsilon > o$ the following inequality holds:

$$(1) \quad \mathbb{E}\, n_t(x+\int_s^t Gd\xi)^2 \leqslant n_s(x)^2 + C(t-s)\|y-z\|^2 + \varepsilon \; ,$$

where $t > s$ and $C = 2\, C(g)^2$. Then the inequality in (ii) follows from the ξ-continuity of g and the continuity property (i) of the family (n_t), if (i) is proved.

We introduce the abbreviations

$$N_t^1(x,X,Y,Z) := \mathbb{E}\|x + \int_t^\infty Z(g(X)-g(Y))d\xi\|^2 \; ,$$

$$N_t^2(x,X,Y,Z) := C \int_t^\infty \mathbb{E}(Z\|X-Y\|)^2 d\lambda \quad \text{and}$$

$$N_t(x,X,Y,Z) := N_t^1(x,X,Y,Z) - N_t^2(x,X,Y,Z)$$

for $x \in F$, $X,Y \in \mathcal{E}_t(E)$ and $Z \in \mathcal{E}_t(\mathbb{R})$. W.l.o.g. we may and do assume that $N_t^2(x,Y,Y,Z) \leqslant k$ for a given constant $k \geqslant o$. The general inequality (1) then follows by a limit argument.

Now let for a given $\delta > o$ (the dependence of ε will be specified later) $(V_j)_{j \geqslant 1}$ denote a partition of F into Borel sets such that for every $j \geqslant 1$ and all $u,v \in V_j$:

$$\|u-v\| < \delta \quad \text{and} \quad |n_t(x+u)^2 - n_t(x+v)^2| < \delta \; .$$

The existence of such a partition follows from the separability of F and the integrability of $n_t(x+\eta)$, where $\eta := \int_s^t G d\xi$.

We choose a fixed $u_j \in V_j$ for every $j \geq 1$. By definition of n_t there exist for every $j \geq 1$ a pair $X_j, Y_j \in \mathcal{E}_t(E)$ and a $Z \in \mathcal{E}_t(\mathbb{R})$ such that

$$n_t(x+u_j)^2 \leq N_t(x+u_j, X_j, Y_j, Z_j)^2 + \delta$$

for every $j \geq 1$. Now we define

$$X := y\, 1_{]s,t] \times \Omega} + \sum_{j \geq 1} X_j\, 1_{[\eta \in V_j]} ,$$

$$Y := z\, 1_{]s,t] \times \Omega} + \sum_{j \geq 1} Y_j\, 1_{[\eta \in V_j]} \text{ and}$$

$$Z := 1_{]s,t] \times \Omega} + \sum_{j \geq 1} Z_j\, 1_{[\eta \in V_j]} .$$

Then we have $X, Y \in \mathcal{E}_s(E)$, $Z \in \mathcal{E}_s(\mathbb{R})$ and

$$\eta + \int_t^\infty Z(g(X)-g(Y))d\xi = \int_s^\infty Z(g(X)-g(Y))d\xi$$

It follows from the ξ-Lipschitz property of g and the choice of the partition (V_j) that there is a constant $K \geq o$ such that

$$(2) \quad \left| \mathbb{E} \|x + \eta + \int_t^\infty Z(g(X)-g(Y))d\xi\|^2 - \mathbb{E}\|x + \zeta + \int_t^\infty Z(g(X)-g(Y))d\xi\|^2 \right| < K\delta ,$$

where $\zeta := \sum_{j \geq 1} u_j 1_{[\eta \in V_j]}$. With the aid of (2) we finally get the following chain of inequalities, which will prove inequality (1):

$$\mathbb{E}\, n_t(x + \int_s^t G d\xi)^2$$

$$\leq \sum_{j \geq 1} P[\eta \in V_j]\, n_t(x + u_j)^2 + \delta$$

$$\leq \sum_{j \geq 1} P[\eta \in V_j]\, N_t(x+u_j, X_j, Y_j, Z_j) + 2\delta$$

$$= \mathbb{E}\|x + \zeta + \int_t^\infty Z(g(X)-g(Y))d\xi\|^2 - c\int_t^\infty \mathbb{E}(Z\|X-Y\|)^2 d\lambda + 2\delta$$

$$\leq \mathbb{E}\|x+\eta+\int_s^\infty Z(g(X)-g(Y))d\xi\|^2 - c\int_t^\infty \mathbb{E}(Z\|X-Y\|)^2 d\lambda + \delta(2+K)$$

$$\leq \mathbb{E}\|x + \int_s^\infty Z(g(X)-g(Y))d\xi\|^2$$
$$- C\int_t^\infty \mathbb{E}(Z\|X-Y\|)^2 d\lambda + \delta(2+K)$$

$$= \mathbb{E}\|x + \int_s^\infty Z(g(X)-g(Y))d\xi\|^2$$
$$- C\int_s^\infty \mathbb{E}(Z\|X-Y\|)^2 d\lambda + C(t-s)\|y-z\|^2 + \delta(2+K)$$

$$\leq n_s(x)^2 + C(t-s)\|y-z\|^2 + \varepsilon , \text{ if } (2+K)\delta \leq \varepsilon.$$

Now let us prove (i). That (n_t) is a family
of norms follows easily from the definition: every
n_t is homogeneous and also convex. Putting $X = Y$
in the definition of n_t as a supremum we get

$$\|x\| \leq n_t(x), \text{ and } n_t(x) \leq \sqrt{2}\|x\| \text{ follows from}$$

$$\mathbb{E}\|x + \int_t^\infty Z(g(X)-g(Y))d\xi\|^2$$
$$\leq 2\|x\|^2 + 2C(g)^2 \int_t^\infty \mathbb{E}(Z\|X-Y\|)^2 d\lambda .$$

The continuity property can be obtained as
follows: For $s \geq 0$, $x \in F$ and every $\varepsilon > 0$ there is a
triple X,Y,Z as above such that

$$n_s(x)^2 \leq \mathbb{E}\|x + \int_s^\infty Z(g(X)-g(Y))d\xi\|^2$$
$$- C\int_s^\infty \mathbb{E}(Z\|X-Y\|)^2 d\lambda + \varepsilon$$
$$= \lim_{t\downarrow s}\left[\mathbb{E}\|x + \int_t^\infty Z(g(X)-g(Y))d\xi\|^2\right.$$
$$\left. - C\int_t^\infty \mathbb{E}(Z\|X-Y\|)^2 d\lambda\right] + \varepsilon$$
$$\leq \lim_{t\downarrow s} n_t(x)^2 + \varepsilon \leq n_s(x)^2 + \varepsilon.$$

3.3 <u>Remarks</u>. (1) If g does not depend on t, then
it is not necessary to consider the whole family
(n_t) of norms. We then take the norm $n := n_0$.
  (2) The inequality in 3.2(ii) can be interpreted as a classical Lipschitz property of g.
Assume for simplicity $g \in Lip_\xi(E,F)$ and define for
every $z \in F$:

$$m(z) := \sup_{y \in F}\left(\mathbb{E}\,n(y+z\xi)^2 - n(y)^2\right)^{1/2} .$$

(where ξ is $N(o,1)$-distributed). Then m is a norm on the subspace $G := \{z \in F : m(z) < \infty\}$ and inequality 3.2(ii) can be written as

$$m\,(g(x)-g(y)) \leqslant C(g)\|x-y\| \quad (x,y \in E).$$

We will use this fact later on.

3.4 <u>Theorem</u>. The following assertions are equivalent:

(1) g is a ξ-Lipschitz function.

(2) There exists a family (n_t) of equivalent norms on F with properties (i) and (ii) of lemma 3.2.

(3) There exists a family (n_t) of equivalent norms on F with property (i) of lemma 3.2 and the following property (ii') instead of (ii):

(ii') there is a constant $D > o$ such that for a given Bernoulli variable ε, all $c \geqslant o$, $t \geqslant o$, $y \in F$, $x_1, x_2 \in E$

$$\mathbb{E}\,n_t(y + c(g(t,x_1)-g(t,x_2))\varepsilon)^2 \leqslant n_t(y)^2 + Dc^2\|x_1-x_2\|^2.$$

<u>Proof</u>: The implication $(1) \Rightarrow (2)$ is given by lemma 3.2. The equivalence of (2) and (3) is a consequence of the following observation: Let x,y be elements of a Banach space $(G, \|.\|)$ and $s > o$. Then the inequalities

(I_1) there exists a constant $C > o$ (independent of x,y and s) such that for all $c \geqslant o$

$$\mathbb{E}\|x + cy\xi\|^2 \leqslant \|x\|^2 + Cc^2s, \text{ and}$$

(I_2) there exists a constant $D > o$ (also independent of x,y and s) such that for all $c \geqslant o$

$$\mathbb{E}\|x + cy\varepsilon\|^2 \leqslant \|x\|^2 + Dc^2s$$

are equivalent: Suppose that (I_2) holds. Then

$$\mathbb{E}\|x + cy\xi\|^2 + \|x\|^2$$
$$= \mathbb{E}\|x + cy\xi\|^2 + \|x - \mathbb{E}(cy\xi)\|^2$$
$$\leqslant \mathbb{E}(\|x+cy\xi\|^2 + \|x-cy\xi\|^2)$$
$$\leqslant \mathbb{E}(2\|x\|^2 + 2Dc^2\xi^2 s) = 2\|x\|^2 + 2Dc^2s$$

and (I_1) holds with $C = 2D$. The proof of $(I_1) \Rightarrow (I_2)$ is similar and is essentially a consequence of the symmetry of ξ .

It remains to prove $(2) \Rightarrow (1)$. We first make the following observation. Let $o \leq s \leq t$ be given, and suppose that

$$X_1, X_2 \in L^2(\Omega, \mathcal{F}_s, P; E), Y \in L^2(\Omega, \mathcal{F}_s, P; F), Z \in L^\infty(\Omega, \mathcal{F}_s, P).$$

Then it follows from (2) (see [7], proof of theorem 2.1) that the following inequality holds:

$$\mathbb{E}\left[n_t(Y + Z(g(s, X_1) - g(s, X_2))(\xi_t - \xi_s))^2 \mid \mathcal{F}_s\right] \tag{I}$$
$$\leq n_s(Y)^2 + CZ^2 \| X_1 - X_2 \|^2 (t-s) \ .$$

Now let $X, Y \in \mathcal{E}(E)$, $Z \in \mathcal{E}(\mathbb{R})$ be given and take an arbitrary $\Delta = (t_k)_{o \leq k \leq n+1} \in \Pi$ such that for every $\omega \in \Omega$ the functions $X.(\omega), Y.(\omega)$ and $Z.(\omega)$ are constant on the intervals $]t_k, t_{k+1}]$ ($o \leq k \leq n$) and vanish outside $[t_o, t_{n+1}]$. Then we obtain with the aid of (I) the inequality

$$\mathbb{E} \| \int Z(g^\Delta(X) - g^\Delta(Y)) d\xi \|^2$$
$$= \mathbb{E} \| \sum_{k=0}^{n} Z_{t_k}(g(t_k, X_{t_k}) - g(t_k, Y_{t_k}))(\xi_{t_{k+1}} - \xi_{t_k}) \|^2$$
$$\leq \mathbb{E}\left(\mathbb{E}\left[n_t\left(\sum_{k=0}^{n} Z_{t_k}(g(t_k, X_{t_k}) - g(t_k, Y_{t_k}))(\xi_{t_{k+1}} - \xi_{t_k})\right) \mid \mathcal{F}_{t_n}\right]\right)$$
$$\leq \mathbb{E}\left(n_{t_n}\left(\sum_{k=0}^{n-1} Z_{t_k}(g(t_k, X_{t_k}) - g(t_k, Y_{t_k}))(\xi_{t_{k+1}} - \xi_{t_k}))^2\right)\right.$$
$$+ C \, \mathbb{E}(Z_{t_n} \| X_{t_n} - Y_{t_n} \|)^2 (t_{n+1} - t_n)$$
$$\leq \ldots \leq C \sum_{k=0}^{n} \mathbb{E}(Z_{t_k} \| X_{t_k} - Y_{t_k} \|)^2 (t_{k+1} - t_k)$$
$$= C \int \mathbb{E}(Z_t \| X_t - Y_t \|)^2 dt \ ,$$

and we get $g \in Lip_\xi(\mathbb{R}_+ \times E, F)$ by the ξ-continuity of g.⌋

3.5 <u>Corollary</u>. For $T \in L(E, F)$ the following assertions are equivalent:

(1) There exists a constant $C \geq 0$ such that for all $X \in L^2(\lambda \otimes P, E)$

$$\mathbb{E} \left\| \int TX \, d\xi \right\|^2 \leq C^2 \int \mathbb{E} \|X\|^2 \, dt$$

(2) Let (ε_n) denote a fixed Bernoulli sequence and put $\mathcal{O}_n := \mathcal{b}(\varepsilon_1, \ldots, \varepsilon_n)$. Then there exists a constant $D \geq 0$ such that for every sequence (X_n) with $X_n \in L^2(\Omega, \mathcal{O}_n, P; E)$ for all $n \geq 1$:

$$\sup_{n \geq 1} \mathbb{E} \left\| \sum_{k=0}^n T X_k \varepsilon_{k+1} \right\|^2 \leq D^2 \sum_{k=0}^n \mathbb{E} \|X_k\|^2 .$$

(3) There exist a Banach space G and operators $S \in L(E, G)$, $R \in L(G, F)$ with $T = R \circ S$, where S has the additional property: there is a constant $K \geq 0$ such that for all $x \in E$ and all $y \in G$

$$\mathbb{E} \| y + Sx \varepsilon \|^2 \leq \|y\|^2 + K^2 \|x\|^2 .$$

Proof. We only have to observe that (1) means $T \in \mathrm{Lip}_\xi(E, F)$. Then the equivalence of (1), (2) and (3) follows from the theorem.

3.6 Definition. (cf.[6] and [7]) An operator $T \in L(E, F)$ is called 2-smooth, if one of the equivalent conditions of corollary 3.5 holds. The operator S in condition (3) of the corollary is called 2-uniformly smooth. $S^2(E, F)$ will denote the space of 2-smooth operators. For $T \in S^2(E, F)$ we denote by $\mathcal{b}_2(T)$ the smallest constant $D \geq 0$, for which the inequality in condition (2) of corollary 3.5 holds. Then $S^2(E, F)$ is a Banach space under the norm $\mathcal{b}_2(.)$. This Banach space will be investigated in more detail in the next paragraph.

3.7 Theorem. The following assertions are equivalent:

(1) $g \in \mathrm{Lip}_\xi(\mathbb{R}_+ \times E, F)$.

(2) For every $t \geq 0$ there exist Banach spaces F_t, G_t, Lipschitz functions $h_t \in \mathrm{Lip}(E, G_t)$, operators $T_t \in L(G_t, F_t)$ which are 2-uniformly smooth, operators $S_t \in L(F_t, F)$ and for $s \leq t$ operators $R_{t,s} \in L(F_s, F_t)$ such that
(i) $\sup_{t \geq 0} \mathcal{b}_2(T_t) < \infty$, $\sup_{t \geq 0} \|S_t\| < \infty$, $\sup_{s \leq t} \|R_{t,s}\| \leq 1$
and

(ii) the following diagram commutes ($o \leqslant s \leqslant t$):

(iii) $g(.,x) \in C_{\xi}(\mathbb{R}_+, F)$ for all $x \in E$.

<u>Proof</u>: The implication $(2) \Longrightarrow (1)$ is proved exactly in the same way as the implication $(2) \Longrightarrow (1)$ of theorem 3.4. So we only prove $(1) \Longrightarrow (2)$. With the notations of theorem 3.4 we put

$$F_t := (F, n_t) \text{ , and } S_t := R_{t,s} := \text{ Identity on } F \text{ .}$$

For every $z \in F$ we define

$$m_t(z) := \sup_{y \in F} \left[\mathbb{E} \, n_t(y + z\epsilon)^2 - n_t(y)^2 \right]^{1/2}$$

and put $G_t := \left\{ z \in F : m_t(z) < \infty \right\}$. Then G_t is a Banach space under the norm m_t, and the injection T_t from G_t into $F_t = F$ is obviously 2-uniformly smooth by the definition of m_t. Finally, it follows from theorem 3.4 that the Lipschitz function $h_t := g(t,.) - g(t,o)$ takes values in G_t, and the theorem is proved.|

If the function g does not depend on t, the theorem gives a simple criterion that g is a ξ-Lipschitz function:

<u>3.8 Corollary</u>. A function $g : E \rightarrow F$ belongs to $Lip_{\xi}(E, F)$ if and only if g is of the form $g = T \circ h + z$ where z is a fixed element in F, h is a "normal" Lipschitz function from E into a third Banach space G, and T is a 2-smooth operator from G to F.

In the case that g is differentiable the property $g \in Lip_{\xi}(\mathbb{R}_+ \times E, F)$ can also be characterized by an analogous property of the derivative g':

<u>3.9 Proposition</u>. Suppose that g is ξ-continuous and continuously differentiable in the second component and denote by g' the derivative $g' : \mathbb{R}_+ \times E \rightarrow L(E,F)$. Then the following assertions are equivalent:

(1) $g \in \text{Lip}_\xi(\mathbb{R}_+ \times E, F)$.

(2) There exists a family $(n_t)_{t \geq 0}$ of equivalent norms on F with property (i) of lemma 3.2 and the following property (ii-) instead of (ii):

(ii-) there is a constant $K \geq 0$ such that for all $t \geq 0$, $y \in F$ and $x, z \in E$

$$\mathbb{E}\, n_t(y + g'(t,x)z\, \varepsilon)^2 \leq n_t(y)^2 + K\|z\|^2 \, .$$

(3) For every $\Delta = (t_k) \in \Pi$ and all $X, Y \in L^2(\lambda \otimes P, E)$

$$\mathbb{E}\,\Big\| \sum_k g'(t_k, X_{t_k}) Y_{t_k} (\xi_{t_{k+1}} - \xi_{t_k}) \Big\|^2$$
$$\leq C \sum_k \mathbb{E}\|Y_{t_k}\|^2 (t_{k+1} - t_k)$$

<u>Proof</u>: To prove (1) $\Rightarrow$ (2), let (n_t) denote the family of norms which is given by the equivalence (1) $\Rightarrow$ (3) of theorem 3.4. The inequality in (ii') implies that for every $t \geq 0$, $h > 0$, $y \in F$ and $x, z \in E$:

$$\mathbb{E}\, n_t(y + h^{-1}[g(t,x) - g(t, x+hz)]\, \varepsilon)^2$$
$$\leq n_t(y)^2 + D\|z\|^2 .$$

Since by assumption g is continuously differentiable we obtain (ii-) with $K := D$ for $h \to 0$.

The implication (2) $\Rightarrow$ (3) is proved exactly in the same way as (3) $\Rightarrow$ (1) of theorem 3.4. So it remains to prove (3) $\Rightarrow$ (1).

Let $\Delta = (t_k)_{1 \leq k \leq n}$ be fixed and $X^1, X^2 \in L^2(\lambda \otimes P, E)$, $Z \in L^\infty(\lambda \otimes P)$ be given. We put $X = X^1$ and $Y = Z(X^2 - X^1)$ and write ξ_k for the increment $\xi_{t_{k+1}} - \xi_{t_k}$. Then the following inequalities hold:

$$\mathbb{E}\,\Big\| \sum_k Z_{t_k} [g(t_k, X^1_{t_k}) - g(t_k, X^2_{t_k})] \xi_k \Big\|^2$$
$$= \mathbb{E}\,\Big\| \sum_k \Big(\int_0^1 g'(t_k, X^1_{t_k} + r(X^2_{t_k} - X^1_{t_k}))dr \Big) \cdot Y_{t_k} \xi_k \Big\|^2$$
$$\leq \mathbb{E}\,\Big(\int_0^1 \Big\| \sum_k g'(t_k, X^1_{t_k} + r(X^2_{t_k} - X^1_{t_k})) \cdot Y_{t_k} \xi_k \Big\|^2 dr \Big)$$
$$\leq C \sum_k \mathbb{E}(Z_{t_k} \|X^2_{t_k} - X^1_{t_k}\|)^2 (t_{k+1} - t_k) \, .$$

This proves that $g \in \text{Lip}_\xi(\mathbb{R}_+ \times E, F)$.

<u>Remark</u>: The proposition shows especially that the family $(g'(t,x))_{t \geq 0, x \in E}$ is contained in $S^2(E, F)$ and uniformly bounded relative to the norm $\mathfrak{S}_2$.

4. EXAMPLES OF 2-SMOOTH OPERATORS

The results of section 3 indicate that the space $S^2(E,F)$ of 2-smooth operators from a Banach space E into a second Banach space F is an interesting space of operators, at least for the problem of stochastic integration of Banach space valued functions. In this section we will give a series of sufficient criteria for operators to be 2-smooth.

A well known subspace of $S^2(E,F)$ is the space $\pi_2(E,F)$ of all 2-absolutely summing operators from E into F. If $T \in \pi_2(E,F)$, and if (X_k) is an arbitrary sequence of E-valued, square integrable random vectors, adapted to the natural filtration $(\mathcal{O}_n)$ defined by a given Bernoulli sequence (ε_n), then $T \in S^2(E,F)$ follows from the inequality

$$\mathbb{E} \left\| \sum_k TX_k \varepsilon_{k+1} \right\|^2 \leqslant \pi_2(T)^2 \sup_{x' \in E', \|x'\| \leqslant 1} \left(\sum_k \mathbb{E} |\langle X_k, x'\rangle|^2 \right)$$

$$\leqslant \pi_2(T)^2 \sum_k \mathbb{E} \| X_k \|^2$$

The following example is also an immediate consequence of the defining inequality (2) of 3.5. Let E_1, F_1 denote further Banach spaces, and suppose that R, S, T are operators with $R \in L(F, F_1)$, $T \in L(E_1, E)$ and $S \in S^2(E,F)$. Then $R \circ S \circ T \in S^2(E_1, F_1)$ and $\acute{o}_2(R \circ S \circ T) \leqslant \|R\| \acute{o}_2(S) \|T\|$. Since a Banach space is 2-smoothable if and only if the identity operator is 2-smooth, we get $L(E,F) = S^2(E,F)$ in the case that E or F are 2-smoothable.

The spaces ℓ_p with $1 \leqslant p < 2$ are simple examples of not 2-smoothable Banach spaces and the next result illustrates the difference between $L(\ell_p)$ and $S^2(\ell_p)(:= S^2(\ell_p, \ell_p))$. Let $L_d(\ell_p)$ denote the space of diagonal operators on ℓ_p (i.e. operators T of the form $Te_i = c_i e_i$, $i = 1, 2, \ldots$, where (e_i) is the unit vector basis in ℓ_p). $L_d(\ell_p)$ can be naturally identified with ℓ_∞. Let $S^2_d(\ell_p)$ denote the space of all 2-smooth diagonal operators. Then of course $S^2_d(\ell_p) = \ell_\infty = L_d(\ell_p)$, if $p \geqslant 2$. For $1 \leqslant p < 2$ we get:

4.1 <u>Proposition.</u> $S^2_d(\ell_p) = \ell_{\frac{2p}{2-p}}$ $(1 \leqslant p < 2)$.

<u>Proof</u>: Let (X,Σ,μ) be a σ-finite measure space and put $E := L^p(X,\Sigma,\mu)$. For an Σ-measurable function g let T_g denote the multiplication operator $T_g f = gf$. If g is bounded, then $T_g \in L(E)$. Now suppose that in addition $g \in L^{\frac{2p}{2-p}}(X,\Sigma,\mu)$, and let (X_k) be an $(\mathcal{O}_k)$-adapted sequence of square-integrable, E-valued random vectors. Then we obtain

$$\mathbb{E}\left\|\sum_k T_g X_k \varepsilon_{k+1}\right\|_p^2 = \mathbb{E}\left(\int\left|\sum_k g X_k \varepsilon_{k+1}\right|^p d\mu\right)^{2/p}$$

$$\leq \left(\int|g|^{\frac{2p}{2-p}} d\mu\right)^{\frac{2-p}{p}} \mathbb{E}\left(\int\left|\sum_k X_k \varepsilon_{k+1}\right|^2 d\mu\right)$$

$$\leq C \sum_k \mathbb{E}\left(\int|X_k|^2 d\mu\right) \quad \text{for} \quad C := \left(\int|g|^{\frac{2p}{2-p}} d\mu\right)^{\frac{2-p}{p}}.$$

For $X = \mathbb{N}$ this inequality implies $\ell_{\frac{2p}{2-p}} C S_d^2(\ell_p)$ for $1 \leq p < 2$ because of $\|\cdot\|_2 \leq \|\cdot\|_p$.

Now let $T = (c_i) \in S_d^2(\ell_p)$ and let (x_k) be an arbitrary sequence in ℓ_2. Define $X_k := x_k e_k$ for all $k \geq 1$. Then $T \in S_d^2(\ell_p)$ implies the inequality

$$\sum_k |c_k|^p |x_k|^p \leq C \left(\sum_k |x_k|^2\right)^{p/2}$$

$$= C \left(\sum_k (|x_k|^p)^{2/p}\right)^{p/2}$$

for a certain constant $C > 0$. Since this inequality is valid for every sequence (x_k) in ℓ_2, it follows that $(|c_i|^p) \in \ell_{q'}$, where q' is the conjugate exponent of $2/p$, i.e. $q' = \frac{2}{2-p}$. This shows $T \in \ell_{\frac{2p}{2-p}}$.

In the following we often make use of the following inequality (I_p) $(1 \leq p < \infty)$, which for a given $T \in S^2(E,F)$ is equivalent to inequality (2) of corollary 3.5 (cf. [6], Remark 3.3).

$$(I_p) \quad \mathbb{E}\left\|\sum_k T X_k \varepsilon_{k+1}\right\|^p \leq C \mathbb{E}\left(\sum_k \|X_k\|^2\right)^{p/2}$$

for every $(\mathcal{O}_k)$-adapted sequence (X_k) of

square integrable E-valued random vectors. It is not difficult to see that there exists a general constant $D_p > 0$ such that (I_p) holds with the constant $C = D_p \, \sigma_2(T)^p$.

Let (E_k) be a sequence of Banach spaces with norms $\|\cdot\|_k$. Then the p-direct sum of (E_k) is defined as the space $(\sum^{\oplus} E_k)_p$ of all sequences (x_k) with $x_k \in E_k$ such that $(\|x_k\|_k) \in \ell_p$ $(1 \leq p \leq \infty)$.

4.2 Proposition. Let (E_k) and (F_k) be two sequences of Banach spaces, and let (T_k) be a sequence of operators $T_k \in S^2(E_k, F_k)$. Define the mapping $T : \prod_k E_k \longrightarrow \prod_k F_k$ by $T(x_k) := (T_k x_k)$ for every $(x_k) \in \prod_k E_k$. Then the following assertions hold:

(1) Let $1 \leq p \leq 2$. Then $T \in S^2((\sum^{\oplus} E_k)_2, (\sum^{\oplus} F_k)_p)$,

   if $(\sigma_2(T_k)) \in \ell_{\frac{2p}{2-p}}$.

(2) Let $2 \leq p \leq \infty$. Then $T \in S^2((\sum^{\oplus} E_k)_p, (\sum^{\oplus} F_k)_2)$

   if $(\sigma_2(T_k)) \in \ell_{\frac{2p}{p-2}}$.

Proof: If $1 \leq p < 2$ and $(\sigma_2(T_k)) \in \ell_{\frac{2p}{2-p}}$, we use inequality (I_p). Let $(X_k) = ((X_{ki})_{i \geq 1})_{k \geq 1}$ be an arbitrary (σ_k)-adapted sequence of square integrable random vectors with values in $E := (\sum^{\oplus} E_k)_2$. Then we get

$$\mathbb{E} \left\| \sum_k TX_k \, \varepsilon_{k+1} \right\|^p = \mathbb{E}\left[\sum_i \left\| \sum_k T_i X_{ki} \, \varepsilon_{k+1} \right\|^p \right]$$

$$\leq \mathbb{E}\left[\sum_i D_p \sigma_2(T_i)^p \left(\sum_k \|X_{ki}\|^2 \right)^{p/2} \right]$$

$$\leq D_p \left(\sum_i \sigma_2(T_i)^{\frac{2p}{2-p}} \right)^{\frac{2-p}{2}} \mathbb{E}\left[\sum_k \left(\sum_i \|X_{ki}\|^2 \right) \right]^{p/2}$$

since $\ell_{\frac{2}{2-p}}$ is the dual of $\ell_{\frac{2}{p}}$. The case $p=2$ is obtained similar.

Now let $2 < p < \infty$, and take a sequence (X_k) of square integrable random vectors in $E = (\sum^{\oplus} E_k)_p$ which is (σ_k)-adapted. Then we get

$$\mathbb{E} \left\| \sum_k TX_k \, \varepsilon_{k+1} \right\|^2 = \mathbb{E}\left[\sum_i \left\| \sum_k T_i X_{ki} \, \varepsilon_{k+1} \right\|^2 \right]$$

$$\leq \mathbb{E}\left[\sum_i \sigma_2(T_i)^2 \sum_k \|X_{ki}\|^2 \right]$$

$$\leq \left(\sum_i \mathfrak{S}_2(T_i)^{\frac{2p}{p-2}}\right)^{\frac{p-2}{p}} \mathbb{E}\left[\sum_k \left(\left(\sum_i \|X_{ki}\|^p\right)^{1/p}\right)^2\right].$$

The cases $p=2$, ∞ are proved similar.

4.3 <u>Corollary</u>. Let (X,Σ,μ) be a $\mathfrak{S}$-finite measure space. Then the following two assertions hold:

(1) If $1 \leq p \leq 2$ and $T \in L^{\frac{2p}{2-p}}(X,\Sigma,\mu;S^2(E,F))$, then
$T \in S^2(L^2(X,\Sigma,\mu;E),L^p(X,\Sigma,\mu;F))$.

(2) If $2 \leq p \leq \infty$ and $T \in L^{\frac{2p}{p-2}}(X,\Sigma,\mu;S^2(E,F))$, then
$T \in S^2(L^p(X,\Sigma,\mu;E),L^2(X,\Sigma,\mu;F))$.

The next result will be used in the last paragraph.

4.4 <u>Proposition</u>. Let (E_k) be a sequence of Banach spaces and let (T_k) be a sequence of operators $T_k \in S^2(E_k,F)$. If $\sum_k \|T_k\|^2 < \infty$, then $T(x_k) := \sum_k T_k x_k$ defines an operator $T \in L((\sum^\oplus E_k)_2,F)$. Suppose that T has the stronger property $\sum_k \mathfrak{S}_2(T_k)^2 < \infty$. Then $T \in S^2((\sum^\oplus E_k)_2,F)$.

<u>Proof</u>: Let $(X_k) = ((X_{ki})_i)_k$ be an arbitrary $(\mathcal{G}_k)$-adapted, square integrable, $(\sum^\oplus E_k)_2$ -valued sequence of random vectors. Then the assertion follows from the following chain of inequalities:

$$\mathbb{E}\left\|\sum_k T X_k \varepsilon_{k+1}\right\|^2 = \mathbb{E}\left\|\sum_i \sum_k T_i X_{ki} \varepsilon_{k+1}\right\|^2$$

$$= \left(\sum_i \mathfrak{S}_2(T_i)^2\right)^2 \mathbb{E}\left\|\sum_i \frac{\mathfrak{S}_2(T_i)^2}{\sum_j \mathfrak{S}_2(T_j)^2} \frac{T_i}{\mathfrak{S}_2(T_i)^2}\sum_k X_{ki}\varepsilon_{k+1}\right\|^2$$

$$\leq \left(\sum_i \mathfrak{S}_2(T_i)^2\right)\sum_i (\mathfrak{S}_2(T_i))^{-2} \mathbb{E}\left\|T_i \sum_k X_{ki}\varepsilon_{k+1}\right\|^2$$

$$\leq \left(\sum_i \mathfrak{S}_2(T_i)^2\right)\sum_i \mathbb{E}\left[\sum_k \|X_{ki}\|^2\right].$$

4.5 <u>Corollary</u>. Let (X,Σ,μ) be a $\mathfrak{S}$-finite measure space and suppose that $T \in L^2(X,\Sigma,\mu;S^2(E,F))$. Then $T \in S^2(L^2(X,\Sigma,\mu;E),F)$.

5.   THE SPACE OF SMOOTHLY INTEGRABLE FUNCTIONS

We know from theorem 2.3 that in general the space $L^2(\lambda \otimes P, E)$ is not contained in $\mathbb{L}^2_\xi$. Motivated by the results of section 3 we introduce in this section a subspace of $L^2(\lambda \otimes P, E)$ which is always contained in $\mathbb{L}^2_\xi$ and which is at the same time rich enough to contain a sufficiently large class of functions, which can occur as integrands of stochastic integrals.

5.1 <u>Definition</u>. A function $f \in L^2(\lambda \otimes P, E)$ is called smoothly <u>integrable</u>, if there exist a Banach space G, a function $g \in L^2(\lambda \otimes P, G)$, and an operator $T \in S^2(G, E)$ such that $f = T \circ g$. The space of all smoothly integrable functions will be denoted by $L^2_s(\lambda \otimes P, E)$.

It follows immediately from the definition that $L^2_s(\lambda \otimes P, E) \subset \mathbb{L}^2_\xi(E)$ and that

$$\mathbb{E} \| \int f d\xi \|^2 \leq \mathscr{C}_2(T)^2 \; \mathbb{E} \int \| g \|^2 \, d\lambda \quad,$$

if (G, T, g) is a triple occuring in definition 5.1 with $T \circ g = f$. Since this factorization of f is surely not unique, we introduce the following notation. For every $f \in L^2_s(\lambda \otimes P, E)$ we denote by $F(f)$ the set of all triples (G, T, g) such that G is a Banach space, $T \in S^2(G, E)$ with $\mathscr{C}_2(T) \leq 1$, and $g \in L^2(\lambda \otimes P, G)$ such that $T \circ g = f$.

For every $f \in L^2_s(\lambda \otimes P, E)$ we define

$$\| f \|_s := \inf_{(G, T, g) \in F(f)} \left(\mathbb{E} \int \| g \|^2 d\lambda \right)^{1/2} .$$

Hence for all $f \in L^2_s(\lambda \otimes P, E)$ we have the inequality

$$\mathbb{E} \| \int f d\xi \|^2 \leq \| f \|^2_s .$$

5.2 <u>Lemma</u>. $\| \cdot \|_s$ is a norm on $L^2_s(\lambda \otimes P, E)$.

<u>Proof</u>: We prove that $\| \cdot \|_s$ is a homogeneous and convex function. Since the homogenity is obvious it remains to prove that $\| \cdot \|_s$ is convex on the set of all $f \in L^2_s(\lambda \otimes P, E)$ with $\| f \|_s \leq 1$.

So let $f_1, f_2 \in L^2_s(\lambda \otimes P, E)$ with $\| f_1 \|_s \leq 1, \| f_2 \|_s \leq 1$ and $\lambda_1, \lambda_2 \in]0, 1[$ with $\lambda_1 + \lambda_2 = 1$ be given. For every $\varepsilon > 0$ there exist triples $(G_1 T_1, g_1) \in F(f_1)$, $(G_2, T_2, g_2) \in F(f_2)$ such that

$$\mathbb{E}\int \| g_1 \|^2 d\lambda \leq \| f_1 \|_s^2 + \varepsilon \quad , \quad \mathbb{E}\int \| g_2 \|^2 d\lambda \leq \| f_2 \|_s^2 + \varepsilon$$

Consider the triple (G, T, g) with $G := (G_1 \oplus G_2)_2$ (the 2-direct sum, cf. section 4), $T := (\sqrt{\lambda_1}T_1, \sqrt{\lambda_2}T_2)$ and $g := (\sqrt{\lambda_1}g_1, \sqrt{\lambda_2}g_2)$. Then we have

$$T g = \lambda_1 f_1 + \lambda_2 f_2 \quad \text{and} \quad \text{by proposition 4.4}$$

$$\mathscr{6}_2(T)^2 \leq \mathscr{6}_2(\sqrt{\lambda_1}T_1)^2 + \mathscr{6}_2(\sqrt{\lambda_2}T_2)^2 \leq 1 .$$

This implies

$$\| \lambda_1 f_1 + \lambda_2 f_2 \|_s^2 \leq \mathbb{E}\int \| g \|^2 d\lambda \; = \; \mathbb{E}\int \| (\sqrt{\lambda_1}g_1, \sqrt{\lambda_2}g_2) \|^2 d\lambda$$

$$= \lambda_1 \, \mathbb{E}\int \| g_1 \|^2 d\lambda \; + \lambda_2 \, \mathbb{E}\int \| g_2 \|^2 d\lambda \leq 1 + \varepsilon ,$$

and hence $\| \lambda_1 f_1 + \lambda_2 f_2 \|_s \leq 1$ since $\varepsilon > 0$ was arbitrary.

5.3 <u>Theorem</u>. $L_s^2(\lambda \otimes P, E)$ is a Banach space relative to the norm $\| . \|_s$.

<u>Proof</u>: We have to prove that every $\| . \|_s$-Cauchy sequence converges to an element in $L_s^2(\lambda \otimes P, E)$. Let $(f_n)_{n \geq 1}$ be a given $\| . \|_s$-Cauchy sequence. By definition of the norm $\| . \|_s$ this means that for every $\varepsilon > 0$ there is an n_ε such that for all $n, m \geq n_\varepsilon$ there exists a triple $(G_{n,m}, T_{n,m}, g_{n,m}) \in F(f_n - f_m)$ such that $\mathbb{E}\int \| g_{n,m} \|^2 d\lambda < \varepsilon$.

From this we get that there exists a strictly increasing sequence (n_k) in $\mathbb{N}$ and a sequence of triples $(G_k, T_k, g_k) \in F(f_{n_{k+1}} - f_{n_k})$ such that

$$\mathbb{E}\int \| g_k \|^2 d\lambda \leq 2^{-2(k+1)} \qquad \text{for all } k \geq 1.$$

Finally, we choose a triple $(G_0, T_0, g_0) \in F(f_{n_1})$

and put for every $k \geqslant 1$: $F_k := G_{k-1}$,

$$S_k := 2^{-k/2} T_{k-1} \quad \text{and} \quad h_k := 2^{k/2} g_{k-1} \ .$$

Then we define

$$F := (\textstyle\sum_{k \geqslant 1}^{\oplus} F_k)_2 \ ,$$

$$S := (S_k)_{k \geqslant 1} \quad \text{and} \quad h := (h_k)_{k \geqslant 1} \ .$$

It follows from proposition 4.4 that

$$\mathfrak{C}_2(S)^2 \leqslant \textstyle\sum_k \mathfrak{C}_2(S_k)^2 = \sum_k 2^{-k} \mathfrak{C}_2(T_{k-1})^2 \leqslant 1 \ .$$

This implies that $(F,S,h) \in F(f)$ for $f := S \circ h$.

By definition of S and f we get for every $k \geqslant 1$:

$$\| f - f_{n_k} \|_s^2 \ = \ \| f - \textstyle\sum_{j=1}^{k-1}(f_{n_{j+1}} - f_{n_j}) - f_{n_1} \|_s^2$$

$$= \ \| S \circ h - \textstyle\sum_{j=1}^{k} S_j h_j \|_s^2 \ = \ \| \textstyle\sum_{j \geqslant k+1} S_j h_j \|_s^2$$

$$\leqslant \textstyle\sum_{j \geqslant k+1} \mathbb{E}\!\int \| h_j \|^2 d\lambda \ = \ \sum_{j \geqslant k} 2^{j+1} \mathbb{E}\!\int \| g_j \|^2 d\lambda$$

$$\leqslant \textstyle\sum_{j \geqslant k} 2^{-(j+1)} \ = \ 2^{-k} .$$

This proves that $\lim\limits_{n \to \infty} \| f - f_n \|_s = 0$.

We close this paragraph in giving a general necessary and sufficient condition that a deterministic function belongs to L_s^2. This criterion may serve as an additional motivation to consider L_s^2 as a best space of functions for which the stochastic integral exists.

5.4 <u>Theorem</u>. For a function $f: \mathbb{R}_+ \longrightarrow E$ the following assertions are equivalent:

(1) $f \in L_s^2(\lambda \otimes P, E)$.

(2) There exists a function $g: \mathbb{R}_+ \longrightarrow \mathbb{R}_+$ such that for all $Z, U \in \mathcal{E}(\mathbb{R})$, $U \geqslant 0$ the following inequality holds:

$$\mathbb{E} \left\| \int Z \, f(U) \, d\xi \right\|^2 \leqslant \int \mathbb{E} \left[Z^2 g(U) \right] d\lambda \ .$$

Proof: If (1) holds, then there exist a Banach space G, a function $h: \mathbb{R}_+ \longrightarrow G$ and a $T \in S^2(G,E)$ such that $f = T \circ h$. Then the inequality in (2) holds with the function $g := \mathfrak{S}_2(T)^2 \|h\|^2$.

Now suppose that (2) holds. We proceed similar as in the proof of lemma 3.2. For all $s \geqslant 0$ we define for all $y \in E$:

$$n_s(y)^2 := \sup_{\substack{Z \in \mathcal{E}_s(\mathbb{R}) \\ U \in \mathcal{E}_s(\mathbb{R}), U \geqslant 0}} \left(\mathbb{E} \left\| y + \int_s^\infty Z f(U) d\xi \right\|^2 \right. \\ \left. -2 \int \mathbb{E} \left[Z^2 g(U) \right] d\lambda \right)$$

Then for every $s \geqslant 0$, n_s is an equivalent norm on E such that $n(y) := n_0(y) = \lim_{s \downarrow 0} n_s(y)$ for all $y \in E$. Exactly as in the proof of lemma 3.2 one can first show that for every $h > 0$, $t \geqslant 0$ the inequality

$$\mathbb{E} \, n_h(y + f(t)\xi_h)^2 \leqslant n(y)^2 + hg(t)$$

holds for every $y \in E$. Dividing this inequality by h we obtain for $h \downarrow o$

$$(*) \quad \mathbb{E} \, n(y + f(t)\xi)^2 \leqslant n(y)^2 + g(t)$$

for all $y \in E$, $t \geqslant 0$, where ξ is $N(o,1)$-distributed. As in the proof of theorem 3.7 we introduce the Banach space G of all $x \in E$ such that

$$m(x) := \sup_{y \in E} \left[\mathbb{E} \, n(y + x\xi)^2 - n(y)^2 \right]^{1/2} < \infty \ .$$

Then the injection T from G into E is 2-smooth. By inequality $(*)$ the function f is G-valued. This proves $f \in L_s^2(\lambda \otimes P, E)$.

Egbert Dettweiler
Mathematisches Institut
der Universität Tübingen
Auf der Morgenstelle 1o
74oo Tübingen
West Germany

REFERENCES

1. Dettweiler, E.: Banach space valued processes with independent increments and stochastic integration. In: Probability in Banach spaces IV, Proceedings, Oberwolfach 1982, Lecture Notes in Math. 99o, 54-83. Berlin-Heidelberg-New York: Springer 1983.
2. Dettweiler, E.: Stochastic Integral Equations and Diffusions on Banach Spaces. To appear in: Probability Theory on Vector Spaces, Proceedings of the third int. Conf., Lublin 1983.
3. Dynkin, E.B.: Markov Processes. Vol.I&II. Berlin-Göttingen-Heidelberg: Springer 1965.
4. Métivier, M.: Semimartingales. Berlin-New York: de Gruyter 1982.
5. Métivier, M., Pellaumail, J.: Stochastic Integration. New York: Academic Press 198o.
6. Pisier, G.: Martingales with values in uniformly convex spaces. Israel J. Math. 2o, 326-35o (1975).
7. Woyczynski, W.A.: Geometry and Martingales in Banach Spaces. In: Winter School on Probability, Karpacz 1975, 229-275. Lecture Notes in Math. 472. Berlin-Heidelberg-New York: Springer 1975.

Akira Ichikawa

A SEMIGROUP MODEL FOR PARABOLIC EQUATIONS WITH
BOUNDARY AND POINTWISE NOISE

1. Introduction.

Recently the semigroup model for controlled partial differential equations and stochastic partial differential equations
has been adopted by many researchers in system theory see [2]
and its references. When control or noise has the so-called
distributed nature, then abstract operators associated with
them are bounded and the system theory can be developed almost as in finite dimensions. However, there are some new
aspects in infinite dimensional systems. One example is a
boundary input (control).

Let O be a bounded open domain in R^d with boundary Γ.
We assume that Γ is a d-1 dimensional variety and that O lies
on one side of Γ. Let $A(x,\partial) = \sum_{|m|\leq 2} a_m \partial^m$ be a uniformly
strongly elliptic operator of order two with smooth coeffi-
cients $a_m(x)$. Consider a nonhomogeneous boundary value problem of mixed type:

$$\partial y(x,t)/\partial t = -A(x,\partial)y(x,t),$$
$$y(x,0) = y_0(x),$$
$$\partial y(x,t)/\partial n + a(x)y(x,t) = g(x)u(t) \quad \text{on} \quad \Gamma,$$

(1.1)

where $y_0(x) \varepsilon L_2(O)$, $g \varepsilon L_2(\Gamma)$, ∂ / ∂_n is the outward normal
derivative, $a(x)$ is a real positive function defined on Γ
and $u(t)$ is a locally integrable function. Define $D(-A)$ as
the closure in $H^2(O)$ of the subspace of $C^2(\bar{O})$ which consists
of functions ϕ satisfying the boundary condition $\partial\phi/\partial n + a\phi$
$= 0$. Let $-A$ be the restriction of $-A(x,\partial)$ on $D(-A)$. Then
$-A$ generates an analytic semigroup $S(t)$ [17]. Let M be the
map: $L_2(\Gamma) \to L_2(O)$ defined by $y = Mg$, where y is the solution of $-A(x,\partial)y = 0$, $\partial y/\partial n + ay = g$. It is known [11], [12],
[15], [16] that (1.1) can be described by the semigroup model

$$y(t) = S(t)y_0 + \int_0^t AS(t-r)Mgu(r)dr.$$

(1.2)

81

L. Arnold and P. Kotelenez (eds.), Stochastic Space-Time Models and Limit Theorems, 81–94.
© 1985 by D. Reidel Publishing Company.

Now we take a feedback input i.e., a function of y,

$$u = f(y) \tag{1.3}$$

where f is a real continuous function on some subspace of $L_2(0)$. Then (1.2) becomes an integral equation

$$y(t) = S(t)y_0 + \int_0^t AS(t-r)Mgf(y(r))dr. \tag{1.4}$$

Then new feature of this model is that the generator A appears in the integral term. The existence and uniqueness problem for systems of type (1.4) has been studied by Lasiecka and Triggiani [12].

Now we replace (1.3) by

$$u = f(y) + kf(y)\dot{w}(t) \tag{1.5}$$

where $k > 0$ and $\dot{w}(t)$ is a white noise. Then (1.4) is formally written

$$y(t) = S(t)y_0 + \int_0^t AS(t-r)Mgf(y(r))dr + k\int_0^t AS(t-r)Mgf(y(r))\dot{w}(r)dr$$

If we replace the last integral by the Ito integral, we obtain

$$y(t) = S(t)y_0 + \int_0^t AS(t-r)Mgf(y(r))dr + k\int_0^t AS(t-r)Mgf(y(r))dw(r), \tag{1.6}$$

where w(t) is the standard Wiener process. We may regard this as an abstract model for a parabolic equation with boundary noise. The stochastic evolution equation of this type is new.

We assume that the fractional powers A^γ, $0 < \gamma < 1$, are well defined. Then $Mg \in D(A^{1-\gamma})$ for any $\gamma > 1/4$ [5], [12], [14]. Now set $z(t) = A^{-\gamma}y(t)$, then (1.6) yields

$$z(t) = S(t)z_0 + \int_0^t S(t-r)bh(z(r))dr + k\int_0^t S(t-r)bh(z(r))dw(r), \tag{1.7}$$

where $z_0 = A^{-\gamma}y_0$, $b = A^{1-\gamma}Mg$ and $h(z) = f(A^\gamma z)$. This is equivalent to (1.6), but it can describe a wider class of systems. So we shall first establish existence, uniqueness

and regularity results for (1.7) and then translates them
to (1.6). It turns out that we can apply our abstract results not only to systems with boundary noise but also to
systems with point noise. Applications will be discussed
in section 3.

Stochastic evolution equations of type (1.7) have been
considered by Ichikawa [8], Da Prato [3] and Da Prato,
Iannelli and Tubaro [4] but not from the point of view of
boundary noise. Curtain [1] has proposed a semigroup model
for systems with boundary noise and it corresponds to the
special case where $f(y)$ is independent of y. Zabczyk [16]
gives a model in which boundary values satisfy stochastic
differential equations. His model falls in the standard
class [2], although the state space is larger.

2. Existence and uniqueness.

Let $S(t)$ be a strongly continuous analytic semigroup on a
real separable Hilbert space Y and $-A$ its infinitesimal
generator. We assume that fractional powers A^Y, $0 < \gamma < 1$
are well defined and that for each $0 < T < \infty$ there exists a
constant $K = K(T) > 0$ such that

$$|A^\gamma S(t)| \leq K/t^\gamma \quad \text{for any} \quad 0 < \gamma < 1 \quad \text{and} \quad t \in (0,T].$$
(2.1)

Here $||$ denotes the norm of bounded linear operators on Y.
We consider the stochastic integral equation

$$z(t) = S(t)z_0 + \sum_{i=1}^{J} \int_0^t S(t-r)b_{0i}h_{0i}(z(r))dr$$

$$+ \sum_{i=1}^{N} \int_0^t S(t-r)b_i h_i(z(r))dw_i(r),$$
(2.2)

where z_0, b_{0i}, $b_i \in Y$, h_{0i} and h_i are Lipschitz continuous
functions on $D(A^\alpha)$ for some $0 < \alpha < 1/2$, J and N are positive integers and $w(t) = (w_i(t))$ is an N-dimensional Wiener
process defined on a probability space (Ω,F,P). We denote
by F_t, the σ-algebra generated by $w(s)$, $0 \leq s \leq t$.
We have the following existence result.

Theorem 2.1. For each $\alpha \leq \beta < 1/2$ there exists a unique
F_t-adapted solution $z(\cdot)$ to (2.2) in $L_p((0,T)\times\Omega;D(A^\beta))$ for

any $2 \leq p < 1/\beta \leq 1/\alpha$. For each $0 \leq \bar{\beta} < \alpha$, the solution $z(\cdot)$ ε $L_p((0,T)\times\Omega;D(A^{\bar{\beta}}))$ for any $2 \leq p < 1/\alpha$. Moreover, if $h_{0i}(0) = h_i(0) = 0$, then

$$E|z(t)|_\beta^p \leq M|z_0|^{p/t^{\beta p}} \qquad \text{a.e.t for some } M = M(p,T) > 0 \quad (2.3)$$

where $||_\beta$ is the graph norm on $D(A^\beta)$.

Proof. For simplicity we shall take $b_{0i} = 0$, $N = 1$ and assume that $h = h_1$ is linear. The general case is an easy modification of this case. Then (2.2) becomes

$$z(t) = S(t)z_0 + \int_0^t S(t-r)bh(z(r))dw(r). \qquad (2.4)$$

Let V be the closed subspace of $L_p((0,T)\times\Omega;D(A^\beta))$ which consists of F_t adapted processes. For each $z(\cdot)$ ε V define $(\Lambda z)(t)$ by the right hand side of (2.4). We shall show that Λ maps V into itself. In fact from (2.1) we have

$$|S(t)z_0|_\beta^p = |A^\beta S(t)z_0|^p \leq K^p|z_0|^p/t^{p\beta} \varepsilon L_1(0,T),$$

$$E\left|\int_0^t S(t-r)bh(z(r))dw(r)\right|_\beta^p \leq C(c_0K|b|)^p\int_0^t [1/(t-r)^{p\beta}]E|z(r)|_\beta^p dr$$

$$\varepsilon L_1(0,T),$$

where c_0 is the norm of f and $C = C(p,T) > 0$ is a constant. Choosing a suitable equivalent norm in $L_p((0,T)\times\Omega;D(A^\beta))$ as in [7], we can show that Λ is a contraction on V. Thus we have a unique solution in V. From (2.4) we have

$$E|z(t)|_\beta^p \leq (2K|z_0|)^p/t^{p\beta} + C(2c_0K|b|)^p\int_0^t \{E|z(r)|_\beta^p/(t-r)^{p\beta}\}dr.$$

Thus by the Gronwall's inequality below [6, p.6] we obtain (2.3). The second part follows similarly from the estimates of type above.

Lemma 2.1.   Let $0 \leq \alpha$, $\beta < 1$, $0 < T < \infty$, $a \geq 0$, $b \geq 0$.   If
u: $[0,T] \to R^1$ is integrable such that

$$0 \leq u(t) \leq at^{-\alpha} + b\int_0^t (t-s)^{-\beta} u(s)ds \quad \text{a.e. } t \in [0,T],$$

then

$$0 \leq u(t) \leq aMt^{-\alpha} \quad \text{a.e. on } [0,T],$$

for some $M = M(b,\alpha,\beta,T) < \infty$.

Now we consider regularity of solutions.

Theorem 2.2.   For each $0 \leq \delta < 1/2$, $z(\cdot) \in C((0,T];L_q(\Omega;D(A^\delta)))$
for any q, $2 \leq q < 1/\delta$ if $\alpha \leq \delta$ and for any $2 \leq q < 1/\alpha$ if
$0 < \delta < \alpha$.
If, in particular, $z_0 \in D(A^\delta)$, $0 \leq \delta < 1/2$, then
$z(\cdot) \in C([0,T];L_q(\Omega;D(A^\delta)))$ for any q with $2 \leq q < 1/\delta$ if
$\alpha \leq \delta$ and $2 \leq q < 1/\alpha$ if $\delta < \alpha$.

Proof.   First we shall show that $E|z(t)|_\delta^q$ is well defined for
each $t > 0$.   We shall only consider the case $\alpha \leq \delta < 1/2$.
In fact we have

$$E\left|\int_0^t S(t-r)bh(z(r))dw(r)\right|_\delta^q \leq C_1\int_0^t \{E|z(r)|_\delta^q/(t-r)^{q\delta}\}dr$$

$$\leq C_1 M|z_0|^q \int_0^t dr/r^{q\delta}(t-r)^{q\delta} \qquad (2.5)$$

$$< \infty \quad \text{for all } t \in (0,T],$$

where C_1 and M are some positive constants.   Hence $E|z(t)|_\delta^q$
is finite for all $t \in (0,T]$.   Now we shall show the continuity
of z(t) in $L_q(\Omega;D(A^\delta))$ for each $t \in (0,T]$.   Let $t > s > 0$
and consider

$$z(t) - z(s) = S(t)z_0 - S(s)z_0 + \int_s^t S(t-r)bh(z(r))dw(r)$$
$$+ \int_0^s [S(t-r) - I]S(s-r)bh(z(r))dw(r).$$

First we show

$$|S(t)z_0 - S(s)z_0|_\delta \to 0 \text{ as } t \downarrow s > 0 \text{ or as } s \uparrow t > 0.$$

Note that

$$S(t)z_0 - S(s)z_0 = \int_s^t AS(r)z_0 dr.$$

Thus

$$|S(t)z_0 - S(s)z_0|_\delta = |A^\delta \int_s^t AS(r)z_0 dr|$$

$$\leq \int_s^t |A^{1+\delta}S(r)z_0| dr$$

$$\leq \int_s^t (\overline{K}/r^{1+\delta})|z_0| dr, \ \overline{K} > 0$$

$$= (\overline{K}/\delta)|z_0|(1/s^\delta - 1/t^\delta)$$

$$\to 0 \text{ as } t \downarrow s > 0 \text{ or as } s \uparrow t > 0.$$

Consider now

$$E|\int_s^t S(t-r)bh(z(r))dw(r)|_\delta^q \leq C_2|z_0|^q \int_s^t [1/r^{q\delta}(t-r)^{q\delta}] dr, \ C_2 > 0,$$

$$\leq (C_2|z_0|^q/(1-q^\delta)s^{q\delta})(t-s)^{1-q\delta}$$

$$\to 0 \text{ as } t \downarrow s > 0 \text{ or } s \uparrow t > 0.$$

Finally we consider

$$E|\int_0^s [S(t-s) - I]S(s-r)bh(z(r))dw(r)|_\delta^q$$

$$\leq C\int_0^s |[S(t-s) - I]S(s-r)b|_\delta^q E|z(r)|_\delta^q dr$$

$$\leq c \int_0^s |\int_0^{t-s} A^{1+\delta} S(u+s-r)b\,du|_\delta^q E|z(r)|_\delta^q dr$$

$$\leq C(|b|/\delta)^q M \int_0^s |1/(s-r)^\delta - 1/(t-r)^\delta|/r^{q\delta} dr$$

$$\rightarrow 0 \text{ as } t \downarrow s > 0 \text{ or as } s \uparrow t > 0.$$

Hence we have shown that

$$E|z(t) - z(s)|_\delta^q \rightarrow 0 \text{ as } t \downarrow s > 0 \text{ or as } s \uparrow t > 0.$$

Now let $z_0 \in D(A^\delta)$. Then clearly $S(t)z_0 \in C([0,T];D(A^\delta))$. We can easily see

$$E|\int_0^t S(t-r)bh(z(r))dw(r)|_\delta^q \rightarrow 0 \text{ as } t \rightarrow 0$$

from the estimate of the form (2.5). Hence $z(\cdot) \in C([0,T]; L_q(\Omega;D(A^\delta)))$.

Remark 2.1. $z(\cdot) \in L_p((0,T)\times\Omega; (A^\beta))$ for any $\beta < 1/2$ and $1 \leq p < 2$. The conclusion of the Theorem 2.2 is also true for any $1 \leq q < 2$.

Remark 2.2. If $z_0 \in D(A^\beta)$ for some $\alpha \leq \beta < 1/2$, then we can develop existence in $C([0,T];L_q(\Omega;D(A^\beta)))$ similar to [4].

Remark 2.3. Existence results in L_p space for the deterministic case is given in [10].

Lastly we consider the continuity of sample paths.

Theorem 2.3. Let $0 \leq \delta < 1/2$. Suppose z_0, $b \in D(A^\delta)$. Then $z(t)$ has a version with continuous sample paths in $D(A^\delta)$.

Proof. Define $z(t) = A^\delta z(t)$, then

$$\overline{z}(t) = S(t)A^{\delta}z_0 + \int_0^t S(t-r)A^{\delta}bh(z(r))dw(r).$$

Thus $z(t)$ is continuous a.s. in $D(A^{\delta})$ if and only if $\overline{z}(t)$ is continuous in Y a.s. But since $A^{\delta}z_0$, $A^{\delta}b \in Y$ and $-A$ is dissipative, by Proposition 3.8 [9] $\overline{z}(t)$ has a version with continuous sample paths in Y.

3. Applications.

3.1. Noise depending on the value at a point. Consider the stochastic parabolic equation

$$dz(x,t) = -A(x,\partial)z(x,t)dt + b(x)z(x_0,t)dw(t), \quad x,x_0 \in O$$
$$z(x,0) = z_0(x), \quad x \in O, . \tag{3.1}$$
$$z(x,t) = 0 \quad \text{on} \quad \Gamma,$$

where O, Γ and $A(x,\partial)$ are given as in section 1. In this case $D(-A) = H^2(O) \cap H_0^1(O)$. Since $H^s(O) \subset C(\overline{O})$ for $s > d/2$ and $D(A^s) \subset H^{2s}(O)$, we have $D(A^{\alpha}) \subset C(\overline{O})$ if $\alpha > d/4$. Hence, if $d \geq 2$, the condition of Theorem 2.1 is not satisfied. So we assumed $d = 1$. Then we can take $\alpha = 1/4 + \varepsilon$ for any $\varepsilon > 0$. And for any $1/4 < \beta < 1/2$ we have a unique F_t-adapted solution $z \in L_p((0,T)\times\Omega;D(A^{\beta}))$ for $2 \leq p < 1/\beta$ to

$$z(t) = S(t)z_0 + \int_0^t S(t-r)bh(z(r))dw(r),$$

where $h(z) = z(x_0)$ is a continuous linear functional on $D(A^{\beta})$, $\beta > 1/4$. We can take $\beta = 1/2 - \varepsilon$ for any small $\varepsilon > 0$. Then $z \in L_{2/(1-\varepsilon)}((0,T)\times\Omega;D(A^{1/2-\varepsilon}))$. More suggestively we write this as $z(\cdot) \in L_{2+}((0,T)\times\Omega;D(A^{1/2-}))$. Similarly we have $z(\cdot) \in L_{4-}((0,T)\times\Omega;D(A^{1/4+}))$. From Theorem 2.2 we have for any $0 \leq \delta < 1/2$, $z(\cdot) \in C((0,T];L_q(\Omega;(A^{\delta}))$ for any q with $2 \leq q < 1/\delta$ if $\delta \geq 1/4 + \varepsilon$, and $2 \leq q < 4$ if $\delta < 1/4 + \varepsilon$. Hence

$$z(\cdot) \; \varepsilon \; C((0,T];L_{2+}(\Omega;D(A^{1/2-}))) \cap C((0,T];L_{4-}(\Omega;D(A^{1/4}))).$$

If, in particular, $z_0 \; \varepsilon \; D(A^\delta)$, $0 \le \delta < 1/2$, then $z(\cdot) \; \varepsilon \; C([0,T];L_q(\Omega,D(A^\delta)))$ for any q with $2 \le q < 1/\delta$ if $\alpha \le \delta$ and $2 \le q < 1/\alpha$ if $\delta < \alpha$.

Since $z_0 \; \varepsilon \; D(A^0) = Y$ we have $z(\cdot) \; \varepsilon \; C([0,T];L_{4-}(\Omega;Y))$.

Summing up we have

$$z(\cdot) \; \varepsilon \; C((0,T];L_{2+}(\Omega;D(A^{1/2-}))) \cap C((0,T];L_{4-}(\Omega;D(A^{1/4+})))$$

$$\cap \; C([0,T];L_{4-}(\Omega;Y))).$$

If z_0, $b \; \varepsilon \; D(A^\delta)$, $0 \le \delta < 1/2$, then from Theorem 2.3 $z(t)$ has a version with continuous sample paths in $D(A^\delta)$. In particular $z(t)$ has alsways a version with continuous sample paths in $Y = L_2(0)$.

3.2. Boundary noise: mixed boundary condition. Consider

$$\partial y(x,t)/\partial t = -A(x,\partial)y(x,t), \; x \; \varepsilon \; 0, \; t > 0,$$
$$y(x,0) = y_0(x), \; x \; \varepsilon \; 0, \qquad\qquad (3.2)$$
$$\partial y(x,t)/\partial n + a(x)y(x,t) = g(x)f(y(\cdot,t))\dot{w}(t), \; x \; \varepsilon \; \Gamma,$$

where $\dot{w}(t)$ is a white noise and f is a real function specified later. As we have seen in section 1, the semigroup model for (3.2) is

$$y(t) = S(t)y_0 + \int_0^t A^\nu S(t-r)A^{1-\nu}Mgf(y(r))dw(r) \qquad (3.3)$$

for $\nu = 1/4 + \varepsilon$, since $Mg \; \varepsilon \; D(A^{3/4-\varepsilon})$ for any $\varepsilon > 0$. The z model corresponding to (3.3) is

$$z(t) = S(t)z_0 + \int_0^t S(t-r)bf(A^\nu z(r))dw(r), \; z_0 = A^{-\nu}y_0 \; \varepsilon \; D(A^\nu).$$

In order to satisfy the condition of Theorem 2.1, we can take

continuous functions f on $D(A^\delta)$ only for $0 \le \delta < 1/4$.
Case(i): $f(y) = c + <f,y>$, where c is a constant and $f \in Y$.
Then we can take $\alpha = \nu = 1/4 + \epsilon$. Hence by Theorems 2.1 and
2.2 we have

$$z(\cdot) \in L_{2+}((0,T)\times\Omega;D(A^{1/2-}))) \cap L_{4-}((0,T)\times\Omega;D(A^{1/4+}))$$

$$\cap \, c((0,T];L_{2+}(\Omega;A^{1/2-})))$$

$$\cap \, c([0,T];L_{4-}(\Omega;(A^{1/4+}))).$$

Thus translating this to $y(t)$ we have

$$y(\cdot) \in L_{2+}((0,T)\times\Omega;D(A^{1/4-})) \cap c((0,T];L_{2+}(\Omega;D(A^{1/4-})))$$

$$\cap \, c([0,T];L_{4-}(\Omega;Y)).$$

Since we cannot expect b in any $D(A^\delta)$, $\delta > 0$, we have sample
continuity of z only in Y. Hence $A^{-\alpha}y(t)$ has continuous sample paths. Let Y_δ be the completion of Y with respect to the
inner product $<<y,z>> = <A^{-\delta}y,A^{-\delta}z>$, $y,z \in Y$. Then $y(\cdot)$ has
continuous sample paths in $Y_{1/4+}$ where $Y_{\delta+}$ means $Y_{\delta+\epsilon}$ for any
$\epsilon > 0$ small.
Case (ii): $f(y) = <f,y>_\Gamma$, $f \in L_2(\Gamma)$. In this case $f(y)$ is
continuous on $D(A^\delta)$ for $\delta > 1/4$. Thus $h(z) = f(A^\nu z)$ is continuous on $D(A^\alpha)$ only for $\alpha > 1/2$. So the assumption of
Theorem 2.1 is not satisfied.

3.3. Boundary noise: Dirichlet boundary condition. Consider

$$\partial y(x,t)/\partial t = -A(x,\partial)y(x,t), \quad x \in O, \; t > 0,$$
$$y(x,0) = y_0(x) \tag{3.4}$$
$$y(x,t) = g(x)f(y(\cdot,t))\dot{w}(t), \quad x \in \Gamma,$$

where O, Γ and $A(x,\partial)$ are given as in section 1, $g \in L_2(\Gamma)$
and f and $w(t)$ are as in (3.3).
    Let D be the Dirichlet operator: $L_2(\Gamma) \to L_2(O)$ defined by

$$y = Dg: \; -A(x,\partial)y(x) = 0, \; y(x)|_\Gamma = 0.$$

Then $Dg \in D(A^{1-\nu})$ for $\nu = 3/4 + \varepsilon$, for small $\varepsilon > 0$. Thus the semigroup model for (3.4) is

$$y(t) = S(t)y_0 + \int_0^t A^\nu S(t-r)A^{1-\nu}Dgf(y(r))dw(r). \qquad (3.5)$$

The z model associated with (3.5) is

$$z(t) = S(t)z_0 + \int_0^t S(t-r)bh(z(r))dw(r), \qquad (3.6)$$

where $b = A^{1-\nu}Dg$, $z_0 = A^{-\nu}y_0 \in D(A^\nu)$ and $h(z) = f(A^\nu z)$.

Hence the condition of Theorem 2.1 is not in general satisfied. However, if we take $f(y) = <f,y>$ with $f \in D(A^{*\delta})$ and $\delta > 1/4$, then $h(z)$ is continuous on $D(A^\alpha)$, $\alpha = \nu - \delta < 1/2$. Thus we can apply Theorems 1 and 2 and obtain solutions for (3.4). We take $\delta = \nu$ so that $h(z)$ is continuous on Y. For example $f \in D(A^*)$. Then

$$z(\cdot) \in L_{2+}((0,T)\times\Omega;D(A^{1/2-})) \cap C((0,T];L_{4-}(\Omega;D(A^{1/4+})))$$

$$\cap C([0,T];L_q(\Omega;Y)) \text{ for any}$$
$$2 \leq q < \infty.$$

Thus $y(\cdot) = A^\nu z(\cdot) \in L_{2+}((0,T)\times\Omega;Y_{1/4+}) \cap C([0,T];L_q(\Omega;Y_{3/4+}))$. Moreover, y has continuous sample paths in $Y_{3/4+}$.

3.4. Pointwise noise. Consider

$$dy(x,t) = -A(x,\partial)y(x,t)dt + \delta(x-x_0)f(y(\cdot,t))dw(t), \quad x_0 \in O,$$
$$y(x,0) = y_0(x), \quad x \in O, \qquad (3.7)$$
$$\partial y(x,t)/\partial n + a(x)y(x,t) = 0, \quad x \in \Gamma,$$

where O, Γ and $A(x,\partial)$ are given as in section 1. Let G be the Green's function defined by

$$-A(x,\partial)G(x,\xi) = \delta(x-\xi), \quad \partial G(x,\xi)/\partial n + a(x)G(x,\xi) = 0, \quad x \in \Gamma.$$

Let $g(x) = G(x,x_0)$. Then the semigroup model for (3.7) is
the following:

$$y(t) = S(t)y_0 + \int_0^t AS(t-r)gf(y(r))dw(r). \qquad (3.8)$$

We assume that A is self-adjoint, then $A^{1-\gamma}g \in Y$ for $\gamma > d/4$.
So we take $d = 1$. Thus the z model for (3.8) is

$$z(t) = S(t)z_0 + \int_0^t S(t-r)bh(z(r))dw(r),$$

where $z_0 = A^{-\alpha}y_0$, $\alpha = 1/4 + \varepsilon$, $b = A^{1-\alpha}g$ and $h(z) = f(A^\alpha z)$.
Hence for any Lipschitz continuous function f on $D(A^\gamma)$,
$0 \le \gamma < 1/4$, We apply Theorems 1 and 2. Let $f(y) = <f,y>$,
$f \in Y$, then we have

$$z(\cdot) \in L_{2+}((0,T)\times\Omega;D(A^{1/2-})) \cap C([0,T];L_{4-}(\Omega;D(A^{1/4+})))$$

and hence

$$y(\cdot) \in L_{2+}((0,T)\times\Omega;D(A^{1/4-})) \cap C([0,T];L_{4-}(\Omega;Y)).$$

Moreover, $y(\cdot)$ has continuous sample paths in $Y_{1/4+}$. A possible choice of A is $-A = d^2/dx^2 - 1$, $D(A) = \{y \in H^2(0,1)$,
$y'(0) = y'(1) = 0\}$. If $d = 2$, we can assume that $h(z)$ is
continuous on $D(A^\gamma)$ for some $\gamma < 1/2$ and obtain solutions as
in subsection 3.3. We may change the mixed boundary condition to the Dirichlet condition. Then for $d \le 2$, we obtain
solutions in a similar manner.

Faculty of Engineering,
Shizuoka University,
Hamamatsu 432, Japan

References

1. Curtain, R.F.: 1979, 'Stochastic distributed systems with point observation and boundary control: An abstract theory, Stochastics', 3, 85-104.
2. Curtain, R.F. and A. J. Pritchard: 1978, Infinite Dimensional Linear Systems Theory, Lecture Notes in Control and Information Sciences 8, Springer-Verlag, Berlin.
3. Da Prato, G.: 1983, 'Some results on linear stochastic evolution equations in Hilbert spaces by semigroup method', Stoch. Anal. Appl., 1, 57-88.
4. Da Prato, G. and M. Iannelli and L. Tubaro: 1979, 'Semilinear stochastic differential equations in Hilbert spaces', Bollettino U.M.I. (5) 16-A, 168-185.
5. Fujiwara, D.: 1967, 'Concrete characterization of the domains of fractional powers of some elliptic differential operators of the second order', Proc. Japan Acad. 43, 82-86.
6. Henry, D.: 1981, Geometric Theory of Semilinear Parabolic Equations, Lecture Notes in Math., 840, Springer-Verlag, Berlin.
7. Ichikawa, A.: 1978, 'Linear stochastic evolution equations in Hilbert space', J. Diff. Eqns., 28, 266-277.
8. Ichikawa, A.: 1980, 'Stability and control of stochastic evolution equations', Analysis and Optimization of Stochastic Systems, O.L.R. Jacobs, et. al. (ed.), Academic Press, London.
9. Ichikawa, A.: 1984, 'Semilinear stochastic evolution equations: Boundedness, stability and invariant measures', Stochastics, 12, 1-39.
10. Ichikawa, A. and A. J. Pritchard: 1979, 'Existence, uniqueness and stability of nonlinear evolution equations', J. Math. Anal. Appl., 68, 454-476.
11. Lasiecka, I.: 1980, 'Unified theory for abstract parabolic boundary problems - A semigroup approach', Appl. Math. Optim., 6, 281-333.
12. Lasiecka, I. and R. Triggiani: 1983, 'Feedback semigroups and cosine operators for boundary feedback parabolic and hyperbolic equations', J. Diff. Eqns., 47, 246-272.
13. Lasiecka, I. and R. Triggiani: 1983, 'Dirichlet boundary control problem for parabolic equations with quadratic cost: Analyticity and Riccati feedback synthesis', SIAM J. Control Optimiz., 21, 41-67.

14. Lions, J.L. and E. Magenes: 1972, <u>Non-Homogeneous Boun-</u>
    <u>dary Value Problems and Applications</u>, I, II, Springer-
    Verlag, Berlin.
15. Triggiani, R.: 1980, 'Well posedness and regularity of
    boundary feedback parabolic systems', <u>J. Diff. Eqns.</u>,
    36, 347-362.
16. Zabczyk, J.: 1978, 'On decomposition of generators',
    <u>SIAM J. Control Optimiz.</u>, 16, 523-534.
17. Yosida, K.: 1978, <u>Functional Analysis</u>, Springer-Verlag,
    Berlin.

P. Kotelenez

ON THE SEMIGROUP APPROACH TO STOCHASTIC
EVOLUTION EQUATIONS

ABSTRACT

This paper deals with properties of mild
solutions of stochastic partial differential
equations (with and without delays). Both the case
of a fixed Hilbert state space and the case of a
nuclear Gel'fand triple as state space are
analyzed. Examples of parabolic and hyperbolic
equations are given.

Contents

L. Arnold and P. Kotelenez (eds.), Stochastic Space-Time Models and Limit Theorems, 95–139.

I. INTRODUCTION

This paper deals with properties of mild solutions
of stochastic evolution equations of the type:

$$dX(t) = [AX(t) + \sum_{i=1}^{m} A_i X(t-i\rho)]dt$$

$$+ dM(t) \qquad , \quad t \geq 0, \ \rho > 0 \qquad (1.1)$$

$$X(t) = F(t) \qquad , \quad t \in [-m\rho, 0]$$

where A generates a strongly continuous semigroup
U(t) (A~U(t)) on a separable Hilbert space H
and the A_i are relatively bounded perturbations of
A . Here, $X_t(s) := X(t+s)$ for $s \in [-m\rho, 0]$, F(t)
is a given initial process, and M is a certain
noise process. Except for the last part of Section
3.1 we shall assume that M is state independent,
i.e. (1.1) is linear.

For state dependent noise on H we refer to
Kotelenez [32], Da Prato [12], and Tudor [50]. If
$A_i \equiv 0$, the coefficients in M(t,x) are
essentially bounded operators and $u(t) \in G(t,\beta_T)$
(cf. the following Def. 1.1) then (1.1) has a unique
mild solution under the same local Lipschitz and
linear growth assumptions as in the case of ordinary
stochastic equations (cf. Kotelenez [32] - A = A(t)
may even depend on t if A(t) generates an
evolution operator U(t,s) satisfying a natural
extension of Def. 1.1 and if the domain of A(t)
Dom(A(t))) is independent of t). If $A_i \equiv 0$ and
U(t) analytic, bilinear equations with unbounded
diffusion coefficients have been solved by DaPrato
[12]. For explicit representations of solutions of
bilinear equations -cf. Balakrishnan [4] and DaPrato,
Iannelli and Tubaro [14]. For bounded A_i , bounded
noise coefficients and an H-valued Wiener process as
driving term (1.1) has been investigated in Tudor [50].
Moreover, for $A_i \equiv 0$ the Markov property for (1.1)
was proved in Arnold, Curtain and Kotelenez [2] under
"natural" assumptions and for a special case the
generator was computed in Curtain [10]. However, the
assumption common to all the results mentioned above
is that M(t,x) is σ-additive on H .
The finitely additive case was

investigated by Dawson [17], [18] and Funaki [19]
(cf. our Section 3.1).

Let us now assume $M(t,x) = M(t)$ state independent
and look at the case $M(t)$ σ-additive and $M(t)$
finitely additive on H separately.

If $M(t)$ σ-additive on H (1.1) is an equation
only on H and we obtain a "nice" mild solution if
$U(t)$ belongs to the class of the following
definition:

Definition 1.1
A strongly continuous semigroup $V(t)$ on a
separable Hilbert space H with Hilbert space norm
$|\cdot|_H$ is called contraction-type or, equivalently,
to be an element of $G(1,\beta_T)$ if for any $T > 0$
there is a $\beta_T < \infty$ and an equivalent Hilbert space
norm $\tilde{|\cdot|}_{H,T}$ s.t.

$$\tilde{\|V(t)\|}_{L(H)} \le e^{\beta_T t} \quad \text{for} \quad 0 \le t \le T ,$$

where the operator norm $\tilde{\|\cdot\|}_{L(H)}$ is computed w.r.t.
$|\cdot|_{H,T}$.

Remark 1.1
1. Examples where the semigroup associated with a
   finite dimensional delay equation is in $G(1,\beta_T)$
   are given in Banks and Kappel [5].
2. Another example is given in Lemma 2.1.2 for $U(t)$
   analytic generated by A with $Dom(A) = Dom(A*)$
   where A* is the adjoint operator.
3. Although $G(1,\beta_T)$ is large there are examples of
   Hilbert spaces and semigroups thereon which are
   not in $G(1,\beta_T)$ (cf. Chernoff [8])

Chapter 2 is devoted to stochastic evolution
equations on the fixed Hilbert space H . Section
2.1 contains maximal inequalities for the stochastic
convolution integral $\int_0^{\cdot} G(\cdot-s)dM(s)$ (which we also
call stochastic *-integral), where $G(t)$ is the
fundamental solution generated by the deterministic
part of (1.1), and it contains regularity properties
following from those inequalities. Moreover, some
regularity results of DaPrato [12] are slightly
generalized and it is shown that his assumption for
maximal regularity of $\int_0^{\cdot} U(\cdot-s)dM(s)$ implies the

coercivity of the generator A of U(t). Section
2.2 gives a weak convergence criterion for (1.1).

In Chapter 3 the results are applied to stochastic
space-time models where M(t) is only finitely
additive on H , e.g., M(t) cylindrical Brownian
motion on H . Consequently, we must redefine (1.1)
on an enlarged space of distributions; it is shown
that this can be done rigorously for a large class
of linear equations. The "correct" state space for
those equations is the strong dual Φ' of a nuclear
Fréchet space and the solutions live on some Hilbert
distribution subspaces of Φ' . For parabolic
equations the case of a bounded domain D in R^n
and the whole $R^n (n \geq 1)$ are studied separately. In
the case of the bounded domain we can describe the
smoothest possible state spaces for the solutions of
our equations. Moreover, nonlinear equations on a
bounded domain are also discussed. Finally, we
define for certain second order equations a
corresponding nuclear set-up. In all cases we give
examples from the literature where these (linear)
processes arise.

Notation and formal assumptions
All stochastic objects are defined on a complete
stochastic basis $(\Omega, F, F_{t \geq 0}, P)$ with right continuous
filtration, and all random variables depending on t
are adapted to F_t . Hilbert spaces will usually be
denoted by $(H, < \cdot, \cdot >_H)$ where $< \cdot, \cdot >_H$ is the scalar
product on H . $L(H,K)$ is the set of bounded
operators from H to K, $L(H) = L(H,H)$ and
$| \cdot |_{L(H)}$ the operator norm.

For stability and invariant measures for (1.1) we
refer to Ichikawa [26] (also for more references)
and for maximum likelihood estimates to Bagchi and
Borkar [3].

2. STOCHASTIC EVOLUTION EQUATIONS

2.1 Maximal Inequalities and Regularity for the
Stochastic *-Integral

The first of our three basic maximal inequalities is
a special case of an inequality proved in Kotelenez
[33] and generalizes the stopped Doob inequality

which Metivier and Pellaumail [40] obtained for martingales (cf. also Metivier [39]). Let c_0 be the universal constant in the Burkholder-Davis-Gundy inequality appearing in the upper estimate for the sup of a real valued martingale by the square root of its quadratic variation (Jacod [29]), and set

$$c_1 := 2(1+c_0^2)$$

$$\hat{\tau} := \begin{cases} \text{ess sup } \tau(\omega) \text{ , } & \text{if it exists} \\ \infty & \text{otherwise} \end{cases} \text{ ,}$$

where τ is a stopping time. Let us make the following assumption

A.2.1.1)
  (i) M is an H-valued square integrable (s.i) cadlag martingale, <M> is its Meyer process and $[M^j]$ is the quadratic variation of its pure jump part (i.e., $M-M^j$ can have only totally inaccessible jumps);
 (ii) for any $T > 0$ $U(t) \in G(1,\beta_T)$ with generator A .

Theorem 2.1.1
Under assumption A.2.1.1):

(I) $\int_0^{\cdot} U(\cdot-s)\,dM(s) \in \begin{cases} \text{continuous a.s. if M is continuous} \\ \text{cadlag \quad a.s. if M is cadlag} \end{cases}$

(II) For any stopping time τ ,

$$E \sup_{0\le t<\tau} \left| \int_0^t U(t-s)\,dM(s) \right|_H^2$$

$$\le c_1 e^{4\beta_{\wedge\hat{\tau}}} E\{(<M>+[M^j])(\tau-)\} \qquad (2.1.1)$$

Remark 2.1.1
  (i) Tubaro [49] has independently proved a similar result to (2.1.1)
 (ii) (2.1.1) obviously implies the weaker inequality: For any $T > 0$ and $\delta > 0$,

$$P\{ \sup_{0\le t\le T} | \int_o^t U(t-s)\,dM(s) |^2_{H_1} \ge \delta \}$$

$$\le c_1 \frac{e^{4\beta_T T}}{\delta^2} E|M(T)|^2_H \qquad (2.1.2)$$

which has been proved in Kotelenez [30] for
evolution operators $U(t,s)$ (cf. Curtain and
Pritchard [11] and Tanabe [47] for the
definition and properties of evolution
operators) instead of semigroups if $U(t,s)$
satisfies an estimate as in Def. 1.1 but
without assuming the existence of a quasi-
generator $A(t)$.

(2.1.2) has been generalized by Tudor [50] to cover
stochastic delay-differential equations. In view of
the application in Theorem 2.1.2 we shall give the
basic maximal inequality with operators slightly
more general than those considered by Tudor [50].
Therefore we make assumption

A.2.1.2.0)
(i) M is an H-valued s.i. cadlag martingale;
(ii) $K(t,s)$, $\tilde{K}(t,s)$, $0\le s\le t$, and $L(t,s,r)$,
     $0\le r\le s\le t$ are families of bounded linear
     operators on H which are strongly continuous
     in t and in s (resp. (s,r)) either strongly
     continuous or finitely valued such that for any
     $T > 0$ there is a constant $\beta_T < \infty$ with:

$$|K(t,s)|_{L(H)} \le e^{\beta_T(t-s)} \qquad 0 \le s \le t \le T$$

$$|L(t,s,r)|_{L(H)} \le \beta_T(t-s) \qquad 0 \le r \le s \le t \le T$$

$$K(s,s) = I \quad \text{(identity operator) for all } s\ge 0$$

$$\tilde{K}(t,r) = K(t,s)\tilde{K}(s,r) + L(t,s,r), \quad 0\le r\le s\le t.$$

From Chojnowska-Michalik [9] we have that
$\int_o^\cdot \tilde{K}(\cdot,s)\,dM(s)$ admits a progressively measurable
version, and by Gihman and Skorohod [21] we may
assume that $|\int_o^\cdot \tilde{K}(\cdot,s)\,dM(s)|_H$ is separable.

Lemma 2.1.1
Under assumption A.2.1.2.0) for any $T > 0$ and all
$\delta > 0$ we have

$$P\{ \sup_{0\leq t\leq T} |\int_0^t \tilde{K}(t,s)\,dM(s)|_H \geq \delta \} \leq \frac{K(T)}{\delta^2}\, E|M(T)|_H^2, \quad (2.1.3)$$

with $\quad K(T) := e^{4\beta_T T} \max\{ (\bar{K}(T))^2, \beta_T \cdot K(T), \beta_T^2 \}$

and $\quad \bar{K}(T) := \sup_{0\leq s\leq t\leq T} |\tilde{K}(t,s)|_{L(H)}$.

The <u>proof</u> is the same as the proof of Theorem 1 in Tudor [50] (cf. Kotelenez [30], Th. 1).

□

Let ρ be a fixed constant from $(0,\infty)$, $A, A_1,\ldots A_m$ linear operators on H, and let us consider the unperturbed delay-differential equation on H :

$$\frac{dX(t)}{dt} = AX(t) + \sum_{i=1}^m A_i X(t-i\rho) \quad\quad (2.1.4)$$

We make the following assumption on (2.1.4):

<u>A.2.1.2)</u>
$A, A_1,\ldots, A_m$ are closed linear operators with dense domains in H s.t.

(i) $\quad A \sim U(t) \in G(1,\beta_T)$ for all $T > 0$;

(ii) for each $i\in\{1,\ldots,m\}$ there is a function $c_i(\cdot)\in L_1[0,\rho]$ with

$\quad |U(t)A_i h|_H \leq c_i(t)|h|_H$ for all $h\in D(A_i), i=1,\ldots,m$

and a.a. $t\in[0,\rho]$.

Obviously, (ii) in A.2.1.2) implies that for each i and a.a. $t\in[0,\rho]$ $U(t)A_i$ can be extended to a bounded operator on H, which we shall also denote by $U(t)A_i$ and

$$|U(t)A_i|_{L(H)} \leq c_i(t) .$$

Moreover, A.2.1.2) implies by Nakagiri [42] the existence of a fundamental solution $G(t)$ for (2.1.4), $t \geq 0$, with the representation

$$G(t) = \sum_{j=1}^\infty U_j(t-(j-1)\rho) \quad\quad (2.1.5)$$

where

$$U_j(t) \equiv 0 \qquad \text{if} \quad t < 0 \quad \text{for all} \quad j$$

$$U_1(t) = U(t) \qquad \text{if} \quad t \geq 0$$

$$U_j(t) = \sum_{i=1}^{(j-1)\wedge m} \int_0^t U(t-s)A_i U_{j-i}(s)\,ds$$

$$\text{if} \quad t \geq 0, \quad j=2,3,\ldots \tag{2.1.6}$$

with "$\wedge$" denoting "min".

From the representation we see that

G(t)   is strongly continuous, bounded,
and $G(0) = I$,

and for $0 \leq r \leq s \leq t$

$$G(t-r) = U(t-s)G(s-r) + R(t,s,r) \tag{2.1.7}$$

with

$$R(t,s,r) := \sum_{k=2}^{\infty} \sum_{i=1}^{(k-1)\wedge m} \int_{s-r-(k-1)\rho}^{t-r-(k-1)\rho} U(t-r-(i-1)\rho-u)$$

$$A_i U_{k-i}(u)\,du \ .$$

Thus, $G(t)$, $U(t)$ and $R(t,s,r)$ satisfy assumption A.2.1.2.0), (ii), on $K(t,s)$, $K(t,s)$ and $L(t,s,r)$. On the other hand, setting for $T > 0$

$$s(p) := \frac{Tj}{2^p} \quad \text{if} \quad s \in \left(\frac{Tj}{2^p}, \frac{T(j+1)}{2^p}\right), \quad j=0,1,\ldots,2^p-1$$

we easily see that

$$\int_0^{\cdot} G(\cdot-s(p))\,dM(s) \quad \text{is} \quad \begin{cases} \text{continuous if } M \text{ is} \\ \qquad\qquad\qquad\quad \text{continuous} \\ \text{cadlag if } M \text{ is cadlag} \end{cases}$$

and for $p \geq q$

$$\widetilde{K}_q^p(t,s) := G(t-s(p)) - G(t-s(q))$$

$$K(t,s) := U(t-s)$$

$$L_q^p(t,s,r) := R(t,s,r(p)) - R(t,s,r(q))$$

also satisfy A.2.1.2.0), ii). As a consequence we obtain (cf. Kotelenez [30], Th. 2):

Theorem 2.1.2
Under assumption A.2.1.2):

(I) $\int_0^{\cdot} G(\cdot-s)dM(s)$ is $\begin{cases} \text{continuous a.s. if } M \text{ is} \\ \qquad\qquad\text{continuous a.s.} \\ \text{cadlag a.s. if } M \text{ is} \\ \qquad\qquad\text{cadlag a.s.} \end{cases}$

(II) For any $T > 0$ there is a constant $K(T) < \infty$
s.t. for all $\delta > 0$

$$P\{ \sup_{0 \le t \le T} |\int_0^t G(t-s)dM(s)|_H \ge \delta \} \le \frac{K(T)}{\delta^2} |M(T)|_H^2 . \qquad (2.1.8)$$

where $K(T)$ is a function of β_T and the $c_i(\cdot)$
in A.2.1.2) as described in Lemma 2.1.1.

We now want to derive a maximal inequality for the
stochastic *-integral w.r.t. a norm stronger than
the norm of the state space of the noise process M.
In order to obtain this result for all $U(t)$ which
are generated by self-adjoint closures A of
strongly elliptic operators on a bounded domain
(and for other operators on R^n - cf. Remark 3.2.4)
we somewhat generalize Dawson's [17] assumptions on
the spectrum of A .

A.2.1.3)
  (i)   M is an H-valued continuous martingale with
        tensor quadratic variation $|M|$ (cf. Metivier
        and Pellaumail [41]) s.t. for all $T > 0$ and
        $\eta > 0$ there is a nuclear positive self-adjoint
        operator Q_η^T on H with:

        $$P\{ \sup_{0 \le s \le t \le T} [M](t) - [M](s) \le Q_\eta^T(t-s) \} \ge 1 - \eta$$

        where for two self-adjoint nonnegative definite
        operators A,B on H : $A \le B \Leftrightarrow B-A$ nonnega-
        tive definite;
  (ii) $A\sim U(t)$, is self-adjoint and has a discrete
        spectrum $\{\lambda_n\}$ s.t. there is a $\gamma_0 < 0$ with

        $$\Sigma(\beta-\lambda_n)^\gamma < \infty \text{ for all } \gamma < \gamma_0.$$

        where $\beta \ge 0$ is chosen s.t. $(A-\beta)$ is negative
        definite.

(iii) $A_1,\ldots,A_m$ as in A.2.1.2), and, moreover, all A_i commute with A .

Remark 2.1.2

(i) Condition (i) in A.2.1.3) is satisfied, if, e.g., $M'(t) = \int_0^t \Phi dW(s)$ for $\Phi \in \Lambda^2(K,H,0,W)$, where W is a K-valued Wiener process with covariance operator Q . In this case,

$$[M]\!](t) = \Phi Q \Phi^*$$

and the existence of Q^T in A.2.1.3) follows from the definition of $\Lambda^2(K,H,0,W)$ (cf. [39])

(ii) Condition (ii) in A.2.1.3) implies the coercivity of -A, whence $U(t) \in G(1,\beta)$ and $U(t)$ analytic. Indeed, since $A \sim U(t)$ there is a $\beta \geq 0$ s.t. $\lambda_n + \beta > 0$ for all n . Set

$$V:=Dom((-A+\beta)^{\frac{1}{2}}), <\cdot,\cdot>_V := <(-A+\beta)^{\frac{1}{2}}\cdot,(-A+\beta)^{\frac{1}{2}}\cdot>_H.$$

Then for $\varphi \in Dom(A)$

$$<-A\varphi,\varphi>_H = <\varphi,\varphi>_V - \beta<\varphi,\varphi>_H .$$

(iii) Condition (i) implies that M has Hölder modulus $\mu = \frac{1}{2}$ in t (where the Hölder modulus is the sup over all possible Hölder exponents - cf. Kotelenez [32]) whence from the analyticity of U(t) and the work of DaPrato, Iannelli and Tubaro [15]

$$\int_0^\cdot U(\cdot-s)dM(s)$$ has Hölder modulus $\mu=\frac{1}{2}$(in t)on H.

Let us define

$$(H_\alpha,<\cdot,\cdot>_\alpha):=(Dom((-A+\beta)^{\frac{\alpha}{2}}),<(-A+\beta)^{\frac{\alpha}{2}}\cdot,(-A+\beta)^{\frac{\alpha}{2}}>_H),\alpha\geq 0$$

where "Dom" denotes the domain of the operator. Hence

$$H = H_0 , \quad V = H_1 .$$

Let "v" denote "max" . In the following theorem G(t) is the fundamental solution of (2.1.4) with representation (2.1.5).

Theorem 2.1.3

Under assumption A.2.1.3):

(I) For any $T > 0$, all $\mu \in [0,\frac{1}{2})$ and all $\alpha \in [0,1)$

$$\int_0^\cdot G(\cdot - s)\, dM(s) \in C^\mu([0,T]; H_0) \cap C([0,T]; H_\alpha) \quad \text{a.s.}$$

(II) For all $T > 0$ and all $\alpha \in [0,1)$ and all
$r > \frac{2|\gamma_0|+1}{1-\alpha} + \frac{1}{2}$ there is a constant
$K = K(T,\alpha,r,\beta,c_i, i=1,\ldots,m) < \infty$ such that for
any $\eta, \delta > 0$

$$P\{\sup_{o \le t \le T} |\int_o^t G(t-s)\,dM(s)|_\alpha \ge \delta\} \le \frac{K}{\delta^{2r}} \sum_n \lambda_n^\gamma [\mathrm{Tr}Q_n^T]^r \quad (2.1.9)$$
$$+ \eta$$

with $\quad \gamma := \dfrac{(r-\frac{1}{2})(1-\alpha)-1}{2}.$

Proof

(i) Let $b(t)$ be a real valued standard Brownian
motion and λ and a some numbers > 0. The following estimate is due to Dawson [17] and basic for
the proof of our theorem:
There is a constant $K < \infty$ s.t. for all sufficiently large $a\lambda$ and for any $T > 0$:

$$P\{\sup_{o \le t \le T} |\int_o^t e^{-\lambda(t-s)}\,db(s)| \ge a^{1/2}\}$$

$$\le K\{(a\lambda)^{\frac{1}{2}}\exp(-a\lambda)[\lambda T + \log((a\lambda)^{\frac{1}{2}})] \quad (2.1.10)$$

$$+(a\lambda)^{\frac{1}{4}}\exp(-\frac{a\lambda}{2})(\log((a\lambda)^{\frac{1}{2}}\exp(-2\lambda T))\}.$$

(ii) Since we may consider $\tilde{\lambda}_n := \lambda_n + \beta$ instead of
λ_n, we may assume w.l.o.g. $0 < \lambda_1 \le \lambda_2 \le \ldots$,
and that there is a CONS $\{\varphi_n\}$ of eigenvectors of
A with λ_n as eigenvalues. Then, we easily see
that $\varphi_n^\alpha := \lambda_n^{-\frac{\alpha}{2}}\varphi_n$, $n \in \mathbb{N}$, is a CONS for H_α,
$\alpha \ge 0$, and for any $x \in H_0$ we set

$$<x,\varphi_n^{\alpha}>_{\alpha} := <x,\varphi_n>_o \lambda_n^{\frac{\alpha}{2}} .$$

(iii) Fix $\alpha \in [0,1)$ and set $C(\eta,T,\mu) = [Tr(Q_\eta^T)]^{-1}$. Denote by μ_n the eigenvalues of Q_η^T and by $[M_n](s)$ the quadratic variation of $<M(s),\varphi_n>_o$. Then for $T > 0$, $\delta > 0$:

$$P\{ \sup_{o \leq t \leq T} |\int_o^t U(t-s)dM(s)|_\alpha^2 \geq \delta^2\}$$

$$\leq \sum_n P\{ \sup_{o \leq t \leq T} (\int_o^t e^{-\lambda_n(t-s)} <dM(s),\varphi_n^{\alpha}>_{\alpha})^2 \geq \delta^2 C(\eta,T,\mu)\mu_n\}$$

$$= \sum_n P\{ \sup_{o \leq t \leq T} (\int_o^t e^{-\lambda_n(t-s)} <dM(s),\varphi_n>_o)^2 \geq \delta^2 C(\eta,T,\mu)\mu_n\lambda_n^{-\alpha}\}$$

$$= \sum_n P\{ \sup_{o \leq t \leq T} (\int_o^t e^{-\lambda_n(t-s)} db([M_n](s)))^2 \geq \delta^2 C(\eta,T,\mu)\mu_n\lambda_n^{-\alpha}\}$$

(since $<M(s),\varphi_u>$ is a $[M_u](s)$ - time changed standard Brownian motion $b(s)$)

$$\leq \sum_n P\{ \sup_{o \leq t \leq T} (\int_o^t e^{-\lambda_n(-s)} db(\mu_n s))^2 \geq \delta^2 C(\eta,T,\mu)\mu_n\lambda_n^{-\alpha}\}$$

by A.2.1.3)(i)

$$= \sum_n P\{ \sup_{o \leq t \leq T} (\int_o^t e^{-\lambda_n(t-s)} db(s))^2 \geq \delta^2 C(\eta,T,\mu)\lambda_n^{-\alpha}\}$$

$$\leq \sum_n K\{([a_n\lambda_n]^{\frac{1}{2}}exp(-a_n\lambda_n)[\lambda_n T + log((a_n\lambda_n)^{\frac{1}{2}})]$$

$$+ (a_n\lambda_n)^{\frac{1}{4}}exp(\frac{-a_n\lambda_n}{2})(log((a_n\lambda_n)^{\frac{1}{2}})^{\frac{1}{2}}exp(-2\lambda_n T)$$

by (2.2.10) with $a_n := \delta^2 C(\eta,T,\mu)\lambda_u^{-\alpha}$ since $a_n\lambda_n \sim \lambda_n^{1-\alpha} \to \infty$, whence for any $r > 0$ there is a $C(K,r)$ s.t. the r.h.s. of the last inequality can be estimated from above by

$$C(K,r)\sum_n \{(a_n\lambda_n)^{\frac{1}{2}}(a_n\lambda_n)^{-r}[\lambda_n T + \log((a_n\lambda_n)^{\frac{1}{2}})]$$

$$+ (a_n\lambda_n)^{\frac{1}{4}}(a_n\lambda_n)^{-\frac{r}{2}}(\log((a_n\lambda_n)^{\frac{1}{2}})^{\frac{1}{2}}\} .$$

If we now choose $r > \dfrac{2|Y_0|+1}{1-\alpha} + \dfrac{1}{2}$ we obtain (2.1.9)
with $U(t)$ instead of $G(t)$. Moreover, by
A.2.1.3)(i) we obtain (2.1.9) for $\int_0^t U(t-s)dM(s+r)$
uniformly in r from an arbitrary fixed interval
$[0,\tilde{T}]$.
(iv) First note that $B \in L(H_0)$ and the fact that
B commutes with A implies that B also commutes
with $(\beta-A)^{\frac{\alpha}{2}}$ (and with $U(t)$), whence B restricted
to H_α is in $L(H_\alpha)$ with $||B||_{L(H_\alpha)} = ||B||_{L(H_0)}$
(cf. Section 3.1) .
From (2.1.7) we obtain

$$\int_0^t G(t-s)dM(s) = \int_0^t U(t-s)dM(s) + \int_0^t R(t,s,s)dM(s)$$

and

$$\int_0^t R(t,s,s)dM(s)$$

$$= \sum_{k=2}^{\infty} \sum_{i=1}^{(k-1)\wedge m} \int_{0}^{t}$$

$$\left[\int_{0}^{t-s-(k-1)\rho} U(t-s-(k-1)\rho-u)A_i U_{k-i}(u)\,du \right] dM(s)$$

$$= \sum_{k=2}^{\infty} \sum_{i=1}^{(k-1)\wedge m} \int_{0}^{t} A_i U_{k-i}(u)$$

$$\left[\int_{0}^{t-u-(k-1)\rho} U(t-u-(k-1)\rho-s)\,dM(s) \right] du$$

(*)

by the stochastic Fubini theorem (Chojnowska-Michalik [9]) and since by assumption A_i and , consequently, also $A_i U_{k-i}(u)$ commute with A and, therefore, with $U(t)$.

On the other hand, A.2.1.2) implies by Nakagiri [42] for any $T > 0$ and $i=1,\dots,(k-1)\wedge m$, $k = 2,3,\dots$ the existence of $\tilde{C}_{i,k,T}(\cdot) \in L_1[0,T]$ s.t.

$$\left| A_i U_{k-i}(u) \right|_{L(H_0)} \le \tilde{C}_{i,k,T}(u) \quad \text{for a.a.} \quad u \in [0,T] \ .$$

But the l.h.s. in the last inequality is equal to $\left| A_i U_{k-i}(u) \right|_{L(H_\alpha)}$. Therefore,

$$\sup_{0 \le t \le T} \left| \int_{0}^{t} R(t,s,s)\,dM(s) \right|_{\alpha}$$

$$\le \sum_{k=2}^{\tilde{T}} \sum_{i=1}^{(k-1)\wedge T} \int_{0}^{T} \tilde{c}_{i,k,T}(u) \cdot \sup_{0 \le t \le T} \left| \int_{0}^{t} U(t-s)\,dM(s+u+(k-1)\rho) \right|_{c}$$

with $\tilde{T}$ the smallest natural number $\ge \frac{T}{\rho} + 1$, since for larger k the integrands in (*) become 0 . Consequently, by step (iii) we obtain (II).
(v) The existence of the spatially smooth version for $\int_{0}^{\cdot} G(\cdot-s)\,dM(s)$ is now an easy consequence by the Borel-Cantelli theorem. By Remark 2.1.2,(iii), $\int_{0}^{\cdot} U(\cdot-s)\,dM(s)$ is in $C^\mu([0,T];H_0)$ a.s. for all $\mu \in [0,\frac{1}{2})$ which implies by the commutativity

assumption w.r.t. A_i, A that

$$\int_0^. R(\cdot,s,s)dM(s) \in C^\mu([0,T];H_0),$$ whence we obtain (I).

□

In [12] DaPrato has obtained spatial regularity
results for the stochastic *-integral by semigroup
methods if the semigroup is analytic. These results
were applied by DaPrato to linear SPDE . Such
regularity results had been obtained earlier by
Pardoux [43] and Krylov and Rozovskij [35] via a
variational approach under a coercitivity assumption
on the generator A (and a monotonicity assumption
if A is nonlinear). In this last part of Section
2.1 we shall extend DaPrato's maximal regularity
results to more general noise terms and to the
fundamental solution G(t) of (2.1.5) assuming in
this latter case that U(t) is analytic. Accordingly,
we assume all Hilbert spaces to be complex for this
last part of Section 2.1. DaPrato's assumption for
maximal regularity is the following:

A.2.1.4)
$A \sim U(t)$ is analytic, and $Dom(A) = Dom(A^*)$,
where A^* is the adjoint operator of A .

Before proceeding further we want to clarify the
relation of assumption A.2.1.4) to the coercivity
assumption which is basic to the variational approach
(cf. Lions and Magencs [38] and Pardoux [43]):

Lemma 2.1.2
 (i) If - A defines a coercive quadratic form, then
U(t) is analytic, and $U(t) \in \overline{G}(1,\beta)$
(ii) Under assumption A.2.1.4) - A defines a coereive
form w.r.t. an equivalent scalar procut $<<\cdot,\cdot>>_0$
on H .

Proof
(i) follows from Tanabe [47], Section 3.6 (A is
called regularly dissipative in [47]).
(ii) DaPrato and Grisvard [13] have shown that under
assumption A.2.1.4) there is an extension $(\hat{A},\hat{U}(t))$
of $(A,U(t))$ onto a Hilbert space $(\hat{H},(\cdot,\cdot))$
$(\hat{H},(\cdot,\cdot)) \supset (H,<\cdot,\cdot>$ (in the sende of continuous
inclusion) s.t. $H = Dom(\hat{A})$, $\hat{U}(t)$ analytic, and

$$|\hat{U}(t)|_{L(\hat{H})} = |U(t)|_{L(H)} \cdot \quad \text{Let us denote by}$$

$$(H_{-\varepsilon}, (\cdot,\cdot)_{-\varepsilon}) \ , \quad \varepsilon \in (0,2)$$

the interpolation spaces between H and $\hat{H}$ in the sense of Butzer and Berens [7] (cf. DaPrato [12]) which are equal to the interpolation spaces of Lions and Magenes [38], $[H,\hat{H}]\varepsilon/2$ (cf. Butzer and Berens [7], Theorems 3.4.2 and 3.5.3). By interpolation we obtain that $(\hat{A},\hat{U}(t))$ has a restriction $(A_{-\varepsilon}, U_{-\varepsilon}(t))$ on $H_{-\varepsilon}$ s.t. $U_{-\varepsilon}(t)$ is analytic (cf. Davies [16], Th. 2.39 and Lions and Magenes [38], Ch. 1, Section 5.1). Now we fix $\varepsilon \in (0,1)$ and interpolate between $H_{-\varepsilon}$ and $D(A_{-\varepsilon})$ using the method of Butzer and Berens (assuming w.l.o.g. that the spectrum of $A \subset (-\infty,0)$):

$$(H_{-\varepsilon+\gamma}, <<\cdot,\cdot>>_{-\varepsilon+\gamma}) \ , \quad \gamma \in (0,2)$$

is the subspace of those $x \in H_{-\varepsilon}$ for which $<<x,x>>_{-\varepsilon+\gamma} =: ||x||^2_{-\varepsilon+\gamma} < \infty$, where

$$<<x,y>>_{-\varepsilon+\gamma} := \int_0^\infty t^{1-\gamma}(A_{-\varepsilon}U_{-\varepsilon}(t)x, A_{-\varepsilon}U_{-\varepsilon}(t)Y)_{-\varepsilon}dt.$$

$$(2.1.11)$$

Note that

$$H_0 = H$$

with equivalent norms. Then set

$$V := H_1 = [D(A),H_0]_{\frac{1}{2}}$$

and after identifying H_0 with its dual H_0^* we obtain

$$V^* = H_{-1} = [H_0,\hat{H}]_{\frac{1}{2}} \cdot$$

from Lions and Magenes [38], Ch. 1, Th. 6.2. Moreover, we have

$$A \in L(D(A),H_0) \quad \text{and} \quad \hat{A} \in L(H_0,\hat{H})$$

which implies by Lions and Magenes [38], Ch. 1,

Th. 5.1, that

$$A_{-1} \in L(V,V^*) .$$

Hence, denoting by $(\cdot,\cdot)_{V^*,V}$ the duality between V^* and V, we obtain for some $c > 0$:

$$|(A_{-1}x_1y)_{V^*,V}| \le c|x|_1 \cdot |y|_1$$

for $x,y \in H_1 = V$.

On the other hand, if $x \in D(A)$ then (for $\varepsilon \in (0,1)$)

$$(A_{-1}x,x) = <<Ax,x>>_0$$

$$= \int_0^\infty t^{1-\varepsilon} (A^2 U(t)x, AU(t)x)_{-\varepsilon} \, dt$$

$$= \int_0^\infty t^{1-\varepsilon} (A\tfrac{\partial}{\partial t}U(t)x, AU(t)x)_{-\varepsilon} dt$$

$$= -\int_0^\infty t^{1-\varepsilon} (AU(t)x, A\tfrac{\partial}{\partial t}U(t))_{-\varepsilon} dt$$

$$- (1-\varepsilon)\int_0^\infty t^{-\varepsilon} |AU(t)x|^2_{-\varepsilon} dt$$

by partial integration, whence

$$\operatorname{Re}(A_{-1}x,x)_{V^*,V} \le \frac{-(1-\varepsilon)}{2}||x||^2_1$$

for $x \in D(A)$ and, consequently, by continuity, also for $x \in H_1 = V$.

$\square$

Corollary 2.1.1

If assumption A.2.1.4) holds, then $U(t) \in G(1,\beta)$.

Instead of A.2.1.4) we may by Lemma 2.1.2 make the following coercivity assumption on A in (2.1.4)

A.2.1.5

V and H are separable Hilbert spaces with

$$V \to H$$

and the topology of V is stronger than that of H. There is a quadratic form $a(x,y)$ on V and constants $c,\mu,\beta > 0$ s.t.

$$|a(x,y)| \leq c|x|_V \cdot |y|_V$$

$$\text{Re } a(x,x) \geq \mu|x|_V^2 - \beta|x|_H^2$$

Remark 2.1.3

We take   A   to be the restriction to   H   of $\hat{A}$
which exists by the Lax-Milgram theorem through

$$(\tilde{A}x,y)_{V*,V} = a(x,y)$$

where V^* is the dual of   V .   Then by Tanabe [47]
both   A   and $\hat{A}$ generate analytic semigroups
$U(t)$ and $\tilde{U}(t)$ on   H   and V^* ,   respectively.
Moreover, $U(t) \in G(1,\beta)$.

In the following Theorem 2.1.4   G   is the fundamenta
solution for (2.1.4) as given in (2.1.5).

Theorem 2.1.4

Let   M   be an H-valued square integrable martingale.
A   is defined in Remark 2.1.3 under assumption
A.2.1.5). For $A_1,\dots,A_m$ assume   A.2.1.2)., (ii)
and A.2.1.3), (ii).
Then

$$\int_0^\cdot G(\cdot-s)dM(s) \in \begin{cases} D([0,\infty);H) \text{ a.s. if M is cadlag a.s.} \\ C^\mu([0,T];H) \text{ a.s. if } M \in C^\mu([0,T];H) \\ \qquad\qquad\qquad \text{a.s., } \mu \in [0,\tfrac{1}{2}), T>0 \\ L_2^F([0,T]\times\Omega;V) \end{cases}$$

Proof

(i) Take $\varepsilon \in (0,\tfrac{1}{2})$ and set

$$H_{-\varepsilon} := [H,V^*]_{\tfrac{1}{2}} .$$

By interpolation we see that $\bar{U}$,   the restriction
of $\tilde{U}(t)$ onto $H_{-\varepsilon}$,   is analytic (cf. proof of
Lemma 2.1.2)
(ii) Let $[\bar{M}]$ denote the quadratic variation of   M
in $H_{-\varepsilon}$ and $\bar{Q}^M$ the "probabilistic" covariance
operator of   M   on $H_{-\varepsilon}$ (i.e.; $\bar{Q}^M$ is the Radon-

Nikodym derivative of the tensor valued measure generated by the tensor quadratic variation over $H_{-\epsilon}$ w.r.t. the Doleans measure generated by $[\overline{M}]$ - cf. Metivier and Pellaumail [41]). $\overline{Tr}$ is the trace on $H_{-\epsilon}$. Then, following DaPrato's proof in [12] (cf. also Metivier and Pellaumail, loc.cit.) we obtain (repeatedly using Fubini's theorem) for $\vartheta = \epsilon + \frac{1}{2}$ that

$$E\int_o^T |\int_o^t U(t-s)dM(s)|_V^2 dt$$

$$= E\int_o^T \int_o^\infty \int_o^t h^{1-2\vartheta}\overline{Tr}(\overline{A}\overline{U}(t-s+k)\overline{Q}_s^M\overline{U}*(t-s+h)\overline{A}*)d[\overline{M}](s)dhdt$$

$$= E\int_o^\infty h^{1-2\vartheta}\int_o^T \int_s^T \overline{Tr}(\overline{A}\overline{U}(t-s+h)\overline{Q}_s^M\overline{U}*(t-s+h)A*)dtd[\overline{M}](s)dh$$

$$\leq E\int_o^\infty h^{1-2\vartheta}\int_o^\infty \int_o^T \overline{Tr}(\overline{A}\overline{U}(p+h)\overline{Q}_s^M\overline{U}*(p+h)\overline{A}*)d[\overline{M}](s)dpdh$$

by change of variable $t = p + s$

$$= E\int_o^\infty \int_o^r (r-p)^{1-2\vartheta}\int_o^T \overline{Tr}(\overline{A}\overline{U}(r)\overline{Q}_s^M\overline{U}*(r)\overline{A}*)d[\overline{M}](s)dpdr$$

by change of variable $h = r - p$

$$= \frac{1}{1-2\epsilon} E\int_o^\infty r^{1-2\epsilon}\int_o^T \overline{Tr}(\overline{A}\overline{U}(r)\overline{Q}_s^M\overline{U}*(r)A*)d[\overline{M}](s)dr$$

$$= \frac{1}{1-2\epsilon} E||M(T)||_o^2 .$$

(iii) From this we obtain as in step (iv) of the proof of Th. 2.1.3 that $\int_o^\cdot G(\cdot-s)dM(s) \in L_2^F([0,T]\times\Omega;V)$.

(iv) The other statements ar obvious in view of Th. 2.1.1 Remark 2.1.2, (iii) and the proof of Th. 2.1.3.

□

Remark 2.1.4
DaPrato [12] has shown by an example that in general
there cannot be found a smaller (smoother) Hilbert
space $K \overset{\subset}{\neq} V$ such that $\int_0^\cdot U(\cdot-s)dM(s)$ lives in
K if M is in H .

Remark 2.1.5
Let W be a K-valued Wiener process, σ an $L(K,H)-$
valued adapted process, and U(t) an analytic
semigroup with generator A .   Then DaPrato has shown
in [12] (without the additional-coercivity-assumption
D(A) = D(A*))   that for all $\alpha \in [0,1)$ and T > 0 :

(i) $\int_0^\cdot U(\cdot-s)\sigma(s)dW(s) \in L_2^F([0,T]\times\Omega; H_\alpha)$

   if $\sigma \in L_2^F([0,T]\times\Omega;L(K,H))$;

(ii) $\int_0^\cdot U(\cdot-s)\sigma(s)dW(s) \in C([0,T]; L^2(\Omega;H_\alpha)$

   if $\sup_i |\sigma e_i|_{C([0,T];L_2(\Omega;H)}$

where $\{e_i\}$ are the eigenvectors of the covariance
operator of W .
Clearly, we can easily obtain the generalizations of
DaPrato's results analogous to Theorem 2.1.4 just
with H_α ($\alpha\in[0,1)$)   instead of $H_1 = V$ and without
the coercivity assumption.

2.2 Weak Convergence for Linear Stochastic Evolution Equations

As an important application of the maximal inequalities of Section 2.1 we obtain weak convergence of
solutions of linear stochastic evolution equations
as a consequence of the weak convergence of their
martingale integrators. E.g., on the basis of (2.1.2)
the following was proved in Kotelenez [34]:
Assume $\{M^n\}_{n\in N\cup\{0\}}$ are H-valued locally s.i.
(l.s.i.) cadlag martingales s.t.

   $M^n \Rightarrow M^0$ on $D([0,\infty);H)$ (converges weakly) and $\{U^n(t,s)\}_{n\in N\cup\{O\}}$ are evolution operators in $G(1,\beta)$ (uniformly w.r.t. β) with quasi-generators $A^n(t)$ s.t.

$$A^n(t) \to A^O(t) \quad \text{suitably.}$$

Then

$$\int_O^\cdot U^n(\cdot,s)dM^n(s) \Rightarrow \int_O^\cdot U^O(\cdot,s)dM^O(s).$$

We can obtain similar results for linear stochastic evolution equations with delay on the basis of (2.1.8) or (2.1.9).

A.2.2.1)

   (I) There are $(A_k,A_{i,k},i=1,\ldots,m)_{k\in N\cup\{O\}}$ satisfying A.2.1.2) with β_T and $c_i(t)$ uniformly in n s.t.

   (i) $\Phi := \overset{\infty}{\underset{n,k=o}{\cap}}$ is a core for A_O, where A_k^n is the n-th power of A_k;

   (ii) $\text{Dom}(A_{i,k}) \supset \text{Dom}(A_k)$ $i = 1,\ldots,m,k \in N\cup\{O\}$;

   (iii) $A_{i,k}$ commute with $A_k,i=1,\ldots,m,k \in N\cup\{O\}$;

   (iv) for all $\varphi \in \Phi$ $A_k\varphi \to A_O\varphi$, $A_{i,k}\varphi \to A_{i,O}\varphi$, $i=1,\ldots,m$, as $k \to \infty$.

   (II) There is a sequence of H-valued l.s.i. cadlag martingales M_k on complete stochastic bases (Ω^k,F^k,F_t^k,P^k) with right continuous filtration, $k \in N\cup\{O\}$,

   s.t. $M_k \Rightarrow M_O$ on $D([O,\infty);H)$.

<u>Theorem 2.2.1</u>
Assume A.2.2.1) and let $G_k(t)$ be the fundamental solutions of (2.1.4) with $A_k, A_{i,u}$ instead of A, A_i.

Then

$$\int_0^\cdot G_k(\cdot-s)dM_k(s) \Rightarrow \int_0^\cdot G_0(\cdot-s)dM_0(s) \text{ on } D([0,\infty);H).$$

Proof

(i) Since $A_{i,k}$ commute with A_k we have on Φ that

$$G_k(t) = U_k(t) + \sum_{i=2}^{l} \sum_{j=1}^{i-1} \sum_{\substack{p_1+\ldots+p_j=l \\ 1\le p_i,\ldots,p_j\le m}} \frac{1}{j!}(t-(i-1)\rho)^j U_k(t-(i-1)\rho)A_{p_1}\ldots A_{p_j}$$

(Nakagiri [42], (3.1.4)).

(ii) Let $\{\varphi_n\}$ be a CONS for H with $\varphi_n \in \Phi$ for all n, denote by $L(\varphi_1,\ldots,\varphi_{n_0})$ the linear span of $\varphi_1,\ldots,\varphi_{n_0}$ and by Π_{n_0} the projection of H onto $L(\varphi_1,\ldots,\varphi_{n_0})$ and set $\Pi_{n_0}^\perp := I - \Pi_{n_0}$ (I is the identity). Then

$$\int_0^t G_k(t-s)dM_k(s) = \int_0^t G_k(t-s)\Pi_{n_0}dM_k(s)$$

$$+ \int_0^t G_k(t-s)\Pi_{n_0}^\perp dM_k(s)$$

$$= \Pi_{n_0}M_k(t) + \int_0^t \frac{\partial}{\partial s}G_k(t-s)(\Pi_{n_0}M_k(s))ds$$

$$+ \int_0^t G_k(t-s)\Pi_{n_0}^\perp dM_k(s)$$

by partial integration.

Step (i) shows how to compute $\frac{\partial}{\partial s}G_k(t-s)(\Pi_{n_0}M_k(s))$ whence by our assumptions and the continuous mapping theorem (cf. Billingsley [6], Ch. I, Th. 5.1) for any n_o we have

$$\int_0^\cdot G_k(\cdot-s)\Pi_{n_o}dM_k(s) \Rightarrow \int_0^\cdot G_o(\cdot-s)\Pi_{n_o}dM_o(s)$$

on $D([0,\infty);H)$. The uniform negligibility of $\int_0^\cdot G_k(\cdot-s)\Pi_{n_o}^\perp dM_k$ follows from (2.1.8) exactly as in Th. 1.1 of Kotelenez [34].

$\square$

It would be easy to give an analogous theorem on weak convergence on $D([0,\infty);H_\alpha)$, $\alpha \in [0,1)$ using (2.1.9).

3. STOCHASTIC SPACE-TIME MODELS

Here we shall consider nuclear spaces Φ^* as state spaces for spatially distributed systems. Φ^* is supposed to be the strong dual of a nuclear Fréchet space Φ (test function space). Since in this case Φ is the projective limit of a sequence of Hilbert spaces (cf. Schaefer [46]) we can anlyze our systems on a scale of Hilbert spaces

$$\Phi\subset...\subset H_\alpha\subset...\subset H=H^*\subset...\subset H_{-\alpha}\subset...\subset\Phi^* ,$$

(3.1)

$$\alpha \in N \text{ or } \alpha \in R_+ ,$$

with dense continous imbeddings s.t. for any $\alpha \in R$ (Z) there is a $\beta > \alpha$, $\beta \in R(Z)$, and

$$H_\beta \rightarrow H_\alpha$$

is nuclear. $H_{-\alpha}$ is the strong dual of H_α (cf. Gel'fand Vilenkin [20]). Typically $H_o = L_2(D)$, $D \subset R^n$, $n \in N$ (s. Sections 3.1-3.3). In order to apply the results of Sections 2.1, 2.2 to this case we must show that if we start with a "nice" semigroup $U(t)$ on H_o, we can extend $U(t)$ to a "nice" semigroup $U_{-\alpha}(t)$ on $H_{-\alpha}$ where $\alpha \geq 0$ depends on the properties of the system. (For the application of 2.3 - see Theorem 3.1.1).

The following definition is motivated by the fact that $U(t) \in G(1,\beta_T)$ was basic for our previous results. (In view of our applications we restrict ourselves to semigroups).

Definition 3.1

A semigroup $U(t) \in G(1,\beta_T)$ with generator A on H_0 is called $G(1,\beta)$-extendible (resp. restrictible) onto $H_{-\alpha}$, $\alpha > 0$, (resp. H_α) if there is a $\beta(T,-\alpha)$ $(\beta(T,\alpha))$ s.t. $U(t)$ can be extended to a bounded operator $U_{-\alpha}(t)$ on $H_{-\alpha}$, $t \geq 0$, (restricted to $U_\alpha(t)$ on H_α) and $U_{-\alpha}(t) \in G(1,\beta(T,-\alpha))$ $(U_\alpha(t) \in G(1,\beta(T,\alpha))$ with generator $A_{-\alpha}$ (A_α) which is an extension (restriction) of A.

Remark 3.1

1. A detailed analysis of the extendibility of analytic semigroups was carried through by DaPrato and Grisvard [13].
2. Clearly, the extensions of $U(t)$, if they exist, are unique.

Remark 3.2

1. (3.1) is in many applications (s. Sections 3.1 - 3.3 below) the natural state space of generalized Gauss processes. Gaussian martingales (independent increments!) were described by Itô [27] on the Schwartz set up

$$S \subset H_\alpha \subset L_2(R^n) \subset H_{-\alpha} \subset S' .$$

Now let us assume we have $U_{-\alpha}(t,s) \in G(1,\beta(T,-\alpha)$ on $H_{-\alpha}$ and a Gaussian martingale M on $H_{-\alpha}$. Then $Y(\cdot) := \int_0^{\cdot} U_{-\alpha}(\cdot,s)dM(s)$ is a continuous Gauss-Markov process on $H_{-\alpha}$ (by Th. 2.1.1 and Arnold, Curtain and Kotelenez [2]). On the other hand, assume the existence of a continuous Gauss-Markov process Y on H_α. Then, by a result of Wittig [56] under a certain relation of the covariance of Y to the covariance of a Gaussian martingale M there is an evolution operator $V_{-\alpha}(t,s)$ on $H_{-\alpha}$ s.t.

$$Y(\cdot) = \int_0^{\cdot} V_{-\alpha}(\cdot,s)dM(s) .$$

2. A more general class of processes with independent increments on a nuclear space has been investigated by Ustunel [51].

In the following sections we shall restrict ourselves to real separable Hilbert spaces H_α .

3.1 Examples of Stochastic Parabolic Equations on a Bounded Domain

Let D be a bounded open subset of R^n, $n \in N$, whose boundary ∂D is a C^∞-hypersurface, which lies on one side of D . Let

$$\overset{\circ}{A} = \sum_{i,j=1}^{d} \partial_i a_{ij} \partial_j + c \qquad\qquad (3.1.1)$$

be a strongly elliptic operator on

$$H_o := L_2(D)$$

with $a_{ij}(=a_{ji})$, $c \in C^\infty(\overline{D})$ (infinitely often differentiable on $\overline{D}$, real valued). We close $\overset{\circ}{A}$ w.r.t. homogeneous Dirichlet boundary conditions and denote this closure of $\overset{\circ}{A}$ by A . Then $-A$ is coercive (cf. Lions and Magenes [38] which can be verified by partial integration of $<-A\varphi,\psi>_o$, φ $\psi \in$ Dom(A). Hence,

  $A\sim U(t)))$ $\in$ $G(1,\beta)$, and $U(t)$ is analytic on H_o, where

$$\beta > \sup_{r \in D} |c(r)|$$

(cf. Tanabe [47], Th. 3.6.1). Moreover, A and U(t) are self-adjoint.
Set

  $C_A^\infty(\overline{D}) := \{\varphi \in C^\infty(\overline{D}) : A^m\varphi|_{\partial D} \equiv O \quad m \in N \cup \{O\}\}$

and

  $(H_m, <\cdot,\cdot>_m) :=$ closure of $C_A^\infty(\overline{D})$ w.r.t. $<\cdot,\cdot>_m$

where

$$<\varphi,\psi>_m := <(\beta-A)^m\varphi,\psi>_o = <(\beta-A)^{\frac{m}{2}}\varphi, (\beta-A)^{\frac{m}{2}}\psi>_o$$

$$\varphi,\psi \in C_A^\infty(\overline{D}) .$$

$<\cdot,\cdot>_m$ is a scalar product ($\beta-A$ is invertible).

For $\alpha \in R_+$ we define H_α by

$$H_\alpha := [H_{m+1}, H_m]_{1-\alpha} \quad \alpha \in [m,m+1],\ m \in N \cup \{0\}\ .$$

which is the intermediate space of Lions and Magenes [38].

$H_\alpha = Dom((\beta-A)^{\frac{\alpha}{2}})$, $\alpha \in R_+$, is a real separable Hilbert space with scalar product

$$<\varphi,\psi>_\alpha := <(\beta-A)^{\frac{\alpha}{2}}\varphi,\ (\beta-A)^{\frac{\alpha}{2}}\psi>_o\ .$$

Now let us recall the definitions of the (standard) Sobolev spaces

$$H^m := (H^m(D); <<\cdot,\cdot>>_m) \quad m \in N \cup \{0\}\ ,$$

which are the spaces of real valued functions, m times differentiable in the generalized sense with all derivatives up to order m square integrable. $<<\cdot,\cdot>>_m$, the standard scalar product on H^m, is defined by

$$<<\varphi,\psi>>_m := \sum_{o \leq |j| \leq m} \int_D \partial^j\varphi\,\partial^j\psi\ ,$$

with $\partial^j = \partial_{j_1}\dots\partial_{j_n}$, $j=(j_1\dots j_n)$, $|j| = j_1+\dots+j_n$

(we shall not explicitly write the integration variable and the integrator if there is no risk of confusion).

For $m=0$ $(H^o, <<\cdot,\cdot>>_o) = (H_o, <\cdot,\cdot>_o)$. Set

$$H^\alpha := [H^{m+1}, H^m]_{1-\alpha},\ \alpha \in [m,m+1],\ m \in N \cup \{0\}.$$

Lemma 3.1.1
For all $\alpha \in R_+$, H_α is a closed subspace of H^α with $<<\cdot,\cdot>>_\alpha$ restricted to H_α being equivalent to $<\cdot,\cdot>_\alpha$.

The proof follows for $\alpha \in N$ from Ladyženskaja [37] and for $\alpha \in R_+ \setminus N$ by interpolation (cf. Kotelenez [34]).

□

Corollary 3.1.1

(I) The imbedding $H_\alpha \to H_\beta$ is Hilbert Schmidt,

$$\alpha > \beta + \frac{n}{2} , \quad \beta \geq 0 . \quad (\text{i.e.}$$

$(i^\alpha_\beta; H_\alpha, H_\beta)$ is an abstract Wiener space)

(II) $H_\alpha \subset C^m(\bar{D})$ whenever $\alpha > m + \frac{n}{2}$, $m \in N \cup \{0\}$.

(III) For arbitrary $m \in N$ there exists an operator P_m s.t.

$$P_m \in L(H_\alpha, H^\alpha(R^n)) \quad , \quad \alpha \leq m$$

$$P_m \varphi = \varphi \text{ Lebesque a.e. on } D \quad \varphi \subset H_o$$

The proof follows from the corresponding statements for H^α (cf. Triebel [48], Lions and Magenes [38]).

□

We set

$$\Phi := \bigcap_{\alpha \geq 0} H_\alpha$$

and endow Φ with the locally convex topology defined by $\langle \cdot, \cdot \rangle_\alpha$, $\alpha \geq 0$. By Cor. 3.1.1, (I), Φ becomes a nuclear Fréchet space with this topology (cf. Schaefer [46]).
Now define

$$\Phi' \text{ to be the strong dual of } \Phi$$

and

$$H_{-\alpha} := \{\varphi' \in \Phi' \mid |\varphi'|_\alpha := \sup_{|\psi|_\alpha \leq 1, \psi \in \Phi} |\varphi'(\psi)| < \infty \}.$$

Remark 3.1.1

(I) Obviously, alle elements φ' from $H_{-\alpha}$ can be extended into continuous linear functions on H_α whence $H_{-\alpha}$ becomes the (distributional) strong dual of H_α . Moreover, we may identify H_o with its strong dual H_o' which implies the chain of (dense) continuous inclusions

$$\Phi \subset H_\alpha \subset H_\gamma \subset H_o = H_o' \subset H_{-\gamma} \subset H_{-\alpha} \subset \Phi' \quad (3.1.2)$$

for $\alpha \geq \gamma \geq 0$ (cf. Gel'fand and Vilenkin [20]).
(II) If $\varphi, \psi \in H_o$, then it is easy to see that

$$<\varphi,\psi>_{-\alpha} = <(\beta-A)^{\frac{-\alpha}{2}}\varphi, (\beta-A)^{\frac{-\alpha}{2}}\psi>_{o}$$

(III) From (I) and Cor. 3.1.1, (I), we obtain

$$(i_{\alpha}^{\gamma}, H_{\gamma}, H_{\alpha}), \quad (i_{\alpha}^{\gamma} \text{ the imbedding of } H_{\gamma} \text{ into } H_{\alpha})$$

is an abstract Wiener space, whenever $\alpha + \frac{n}{2} < \gamma, \alpha, \gamma$
$\in \mathbb{R}$ (cf. Kuo [36]). Indeed, by Cor. 3.1.1 there is
a CONS of eigenvectors $\{\varphi_1\} \subset C_A^\infty(\bar{D})$ for $(\beta-A)$

with eigenvalues $\lambda_\ell > 0$ Hence, $\varphi_1^\gamma := \varphi_1 \lambda_1^{-\gamma/2}$ is

a CONS for H_γ, $\gamma \in R$, implying $\sum_\ell |\varphi_\ell^\gamma|_\alpha^2 = \sum_\ell |\varphi_\ell^{\gamma-\alpha}|_o^2 < \infty$

iff $\gamma > \frac{n}{2} + \alpha$. Moreover, we have

$$H_\gamma = \{\varphi' \in \Phi' : \Sigma[\varphi'(\varphi_1)]^2 \lambda_1^\gamma < \infty \} .$$

(IV) DaPrato and Grisvard [13] have given a concrete
representation of $H_{-\alpha}$ as the quotient space
$H_o^2 / G((\beta-A)^{\frac{\alpha}{2}})$, where $G((\beta-A)^{\frac{\alpha}{2}})$ is the graph of
$(\beta-A)^{\frac{\alpha}{2}}$.

The following lemma shows that $U(t)$ and A have
nice extensions (resp. restrictions) to H_α, $\alpha \in R$.

Lemma 3.1.2
For any $\alpha < 0$) $U(t)$ and A have extensions
(resp. restrictions) $U_\alpha(t)$ and A_α on H_α such
that:

$A_\alpha \sim U_\alpha(t) \in G(1,\beta)$, $Dom(A_\alpha) = H_{\alpha+2}$, and $U_\alpha(t)$ is
analytic.

The proof is given in Kotelenez [34].

                                                        □

Let M be an $H_{-\alpha}$-valued l.s.i. cadlag martingale
($\alpha \geq 0$) and consider the stochastic evolution
equation:

$$\begin{aligned} dY(t) &= A_{-\alpha}Y(t)dt + dM(t) \\ Y(o) &\in H_{-\alpha} \end{aligned} \right\} \quad (3.13)$$

As a consequence of Lemma 3.1.2 and Th. 2.1.4 (resp. Th. 2.1.3) we obtain:

Corollary 3.1.2

(3.1.3) has a (unique) mild solution

$$Y(\cdot) = U_{-\alpha}(\cdot)Y_o + \int_o^{\overset{\cdot}{}} U_{-\alpha}(\cdot-s)dM(s)$$

s.t.

$$Y \in \begin{cases} L_2^F([0,T]\times\Omega_i;H_{-\alpha+1}) & \text{if } M \text{ square integrable,} \\ D([0,\infty);H_{-\alpha}) \text{ a.s. it } M \text{ is cadlag} & (3.1.4) \\ C^\mu([0,T];H_{-\alpha}) \cap C((0,T];H_{-\alpha+\epsilon}) \text{ a.s.} \\ \qquad \text{for all } T>0, \ \mu \in [0,\tfrac{1}{2}), \ \epsilon \in [0,1) \\ \qquad\qquad \text{if } M \text{ satisfies } A.2.1.3) \end{cases}$$

Example 3.1.1

Let $a_{ij} = 0$ if $i \neq j$ in (3.1.1) $c = b - d$, where $b,d \in C^\infty(\overline{D},R_+)$ and $c_o \in C^\infty(\overline{D},R_+)$. Moreover, let $X_o \in C^{3+\mu}(\overline{D},R_+)$ ($\mu \in (0,1)$ and assume

$$X_o|_{\partial D} \equiv 0 \ , \quad \overset{o}{A}X_o + c_o|_{\partial D} \equiv 0 \ , \quad \overset{o}{A}^2 X_o + c_o|_{\partial D} \equiv 0 \ .$$

Under these assumptions there is a unique positive solution X to the deterministic PDE

$$\left. \begin{aligned} \tfrac{\partial}{\partial t}X(t,r) &= AX(t,r) + c_o(r) \\ X(o) &= X_o \end{aligned} \right\} \quad (3.1.5)$$

which can be obtained as the high density limit (or thermodynamic limit) of a system branching diffusions (with birth rate $b(r)$ death rate $d(r)$, immigration $c_o(r)$ and emigration (or leaking) due to the boundary conditions) (Kotelenez [34]).

Set

$$F_L(X(t)) := \sum_{i=1}^n -2\partial_i X(t)a_i\partial_i$$

$$F_R(X(t)) := (b+d)X(t) + c_o$$

and let M_J be the (independent) Gaussian martingales with characteristic functional

$$\exp(\frac{1}{2}\int_0^t <F_J(x(s)\varphi,\varphi>_0 ds), \quad \varphi \in \Phi, \quad J \in \{L,R\}.$$

Then, by Itô [27], M_L lives on $H_{-\alpha}$ and M_R on $H_{-\alpha+1}$ for all $\alpha > \frac{n}{2} + 1$. Setting

$$M = M_L + M_R$$

in (3.13) we obtain the fluctuation process Y for the branching diffusions around X as the mild solution to (3.1.3). Moreover, in this case

$$Y \in C^\mu([0,T];H_{-\alpha}) \cap C((0,T];H_{-\alpha+1}) \quad \text{a.s.} \quad (3.1.6)$$

for all $\alpha > \frac{n}{2} + 1$ (cf. Kotelenez [34]).

Example 3.1.2
A similar result was obtained in Kotelenez [31] for $D = [0,1]$ with Neumann boundary conditions.

Example 3.1.3
Let $D = [0,L]$, $L > 0$, $\overset{\circ}{A} = \frac{\partial^2}{\partial x^2} - 1$, and let M be cylindrical Brownian motion on $L_2(0,L)$. Clearly, M is a Gaussian martingale on $H_{-\alpha}$ for all $\alpha > \frac{1}{2}$. The mild solution Y to (3.1.3) was obtained by Walsh [52] as the diffusion approximation to a stochastic neural response model. In particular, (3.1.4) shows that

$$Y \in C((0,T];H_\varepsilon) \quad \text{a.s.} \quad \text{for all} \quad \varepsilon < \frac{1}{2},$$

i.e. Y is function valued. However, by using multi-parameter estimates Walsh (loc. cit) has shown that as a function of $(t,r) \in [0,T] \times [0,L]$ has a modulus of continuity of the order of $\delta^{\frac{1}{4}}(\log\delta^{-1})^{\frac{1}{2}}$

with $\delta = \{(t-s)^2 + (r-q)^2\}^{\frac{1}{2}}$. In particular, $Y(t,r)$ has Hölder modulus $\frac{1}{4}$ on $[0,T] \times [0,L]$ for all $T > 0$.

Remark 3.1.1

We can completely analogously construct the nuclear space (3.1.2) associated with an arbitrary self-adjoint strongly elliptic operator $\tilde{A}$ of order $2m$, $m \geq 1$, and show the extendibility (resp. restrictibility) of $\tilde{A}$ and $\tilde{U}(t)$ (the semigroup generated by $\tilde{A}$) to $\tilde{A}_\gamma$ and $\tilde{U}_\gamma(t)$ onto $\tilde{H}_\gamma$ where $\tilde{H}_\alpha = \text{Dom}(\beta-A)^{\frac{\alpha}{2m}}$. Let M be an $H_{-\alpha}$-valued martingale satisfying A.2.1.3). Then Th. 2.1.3 yields that

$$
\left.
\begin{aligned}
d\tilde{Y}(t) &= \tilde{A}_{-\alpha}\tilde{Y}(t)dt + dM(t) \\[2mm]
\tilde{Y}(0) &\in H_{-\alpha}
\end{aligned}
\right\} \quad (3.1.7)
$$

has a unique mild solution $Y(\cdot) =$

$$\tilde{U}_{-\alpha}(\cdot)\tilde{Y}_0 + \int_0^\cdot \tilde{U}_{-\alpha}(\cdot-s)dM(s) \in C^\mu([0,T];H_{-\alpha}) \cap C((0,T];\tilde{H}_{-\alpha+\varepsilon})$$

for all $\mu \in [0,\frac{1}{2})$, $\varepsilon \in [0,m)$

(cf. Agmon [1]). Thus, if $2m > n$ and M lives on $\tilde{H}_{-\alpha}$ for all $\alpha > \frac{n}{2}$, then

$$\tilde{Y} \in C([0,T];\tilde{H}_\varepsilon) \quad \text{a.s.} \quad \text{for all} \quad \varepsilon \in [0,m-\frac{n}{2})$$

i.e. $\tilde{Y}$ is function valued. In particular, M is a cylindrical Brownian motion B on $\tilde{H}_0$ (since $(i_{-\alpha}^0, \tilde{H}_0, \tilde{H}_{-\alpha})$ is an abstract Wiener space for $\alpha > \frac{n}{2}$) . In this case one can also consider the nonlinear stochastic evolution equation

$$
\left.
\begin{aligned}
d\tilde{Y}(t) &= \tilde{A}_{-\alpha}\tilde{Y}(t)dt + B(t,\tilde{y}))dt + C(t,\tilde{y}(t))dB(t) \\[2mm]
\tilde{Y}(0) &\in H_0
\end{aligned}
\right\} (3.1.8)
$$

for certain B and C . This was done by Dawson [17], [18], and Funaki [19]. Let us look only at Funaki's results and assume

A.3.1.1)

For arbitrary $T > 0$

(i) $C(\cdot,\cdot)$: $[0,T] \times H_0 \longrightarrow L(H_0)$ s.t.

(i.1) $C^*(\cdot,\cdot)\varphi_\ell \in C([0,T]xH_o;H_o)$ $\forall\ell$, where $\{\varphi_\ell\}$
       are the eigenvectors of $\tilde{A}$ and C^* is the
       adjoint of $C(t,x)$.

Moreover, there is a $K < \infty$ s.t.

(i.2) $|C(t,x)|_{L(H_o)} \leq K$

(i.3) $|(C^*(t,x)-C^*(t,y))\varphi_\ell|_o \leq K|x-y|_o$ $\forall$ $x,y\in H_o$, $\ell\in N$

(ii) $B(\cdot,\cdot) \in C([0,T]xH_o;H_o)$ and

(ii.2) $|B(t,x) - B(t,y)|_o \leq K|x-y|_o$ $\forall x,y \in H_o$.

(ii.3) $|B(t,x|_o \leq K$ $\forall x \in H_o$

Under A.3.1.1) Funaki [19] has shown that (3.1.8)
has a unique solution Y which is jointly continuous
if $\tilde{Y}_0(r)$ is continuous. The key to this result is
the estimate for $X(t) := \tilde{Y}(t) - \tilde{U}(t)\tilde{Y}_o$:

For any (large) $p > 0$ there is a $C < \infty$ s.t.

$$E(|X(t,r)-X(s,q)|^{2p}) \leq C\{|r-q|^\gamma+|t-s|^{1-\frac{n}{2m}}\}^p \qquad (3.1.9)$$

with $0 < \gamma < 2m-n$.

However, using Ibragimov's extension of Kolmogorov's
theorem to the multiparameter case (cf. [25], Th. 6)
(3.1.9) implies the even better

Theorem 3.1.1
Assume A.3.1.1) and that $\tilde{Y}(o,r)$ has Hölder
modules $\frac{1}{4}$ on $\overline{D}$. Then, for any $T > 0$

   $\tilde{Y}(t,r)$ is jointly Hölder continuous on $[0,T]x\overline{D}$
for all Hölder exponents $\alpha < \frac{2m-n}{4m}$.

Remark 3.1.2
Th. 3.1.1 implies in particular for $m = n = 1$ that
the solution of the nonlinear SPDE (3.1.8) has the
same Hölder modulus on $[0,T]x\overline{D}$ as the solution of
Walsh's linear SPDE (cf. Ex. 3.1.3).

3.2 Examples of Stochastic Parabolic Equations on R^n

The standard nuclear Gel'fand triple for this case
is the Schwartz triple

$$S(R^n) \subset L_2(R^n) \triangleq L_2'(R^n) \subset S'(R^n) \qquad (3.2.1)$$

A description of (3.2.1) with "intermediate" Hilbert
spaces H_α can be found in Holley and Stroock [24].
Let us briefly recall how the H_α are generated.
Set

$$f_k(x) := (-1)^k e^{\frac{x^2}{2}} \partial_x^k e^{-x^2}, \quad x \in R$$

and

$$g_k(x) := (\pi^{\frac{1}{2}} 2^k k!)^{-\frac{1}{2}} f_k(x) \quad x \in R$$

g_k is called the k-th Hermite function and $\{g_k\}$
are a CONS in $L_2(R)$. Set

$$h_k(x) := g_{k_1}(x_1) \cdot \ldots \cdot g_{k_n}(x_n) , \quad x \in R^n \qquad (*)$$

$$x = (x_1, \ldots, x_n) \in R^n, \quad k = (k, \ldots, k_n) \qquad (N \cup \{0\})^n.$$

Then, $\{h_k\}$ is a CONS in $H_o := L_2(R^n)$ and all
$h_k \in S := S(R^n)$. Moreover, if $|x|^2$ denotes the
multiplication operator $(x \in R^n, |\cdot|$ Euclidean
norm) and Δ the Laplacian on R^n then

$$(-\Delta + |x|^2) h_k = (2|k| + n) h_k , \qquad (3.2.2)$$

where $|k| = k_1 + \ldots + k_n$. Thus $(-\Delta + |x|^2)$ plays the
role of $(\beta - A)$ of Section 3.1 and (omitting the
R^n in $S(R^n)$ etc.) we obtain

$$S \subset H_\alpha \subset H_\gamma \subset H_o = H_o' \subset H_{-\gamma} \subset H_{-\alpha} \subset S' \qquad (3.2.3)$$

for $\alpha \geq \gamma \geq 0$ where

$$H_\alpha := \{\varphi \in S' : |\varphi|_\alpha := |(-\Delta + |x|^2)^{\frac{\alpha}{2}} \varphi|_o < \infty\}, \quad \alpha \in R$$

From (3.2.2) we obtain

Remark 3.2.1

The imbedding $H_\alpha \longrightarrow H_\beta$ is Hilbert Schmidt, whenever $\alpha > \beta + n$ (i.e. $(i_\beta^\alpha, H_\alpha, H_\beta)$ is an abstract Wiener space).

In view of the following applications we would like to solve the analogue of (3.1.3) with $\frac{1}{2}\Delta$ instead of A . Therefore, denoting by $S(t)$ the semigroup generated by $\frac{1}{2}\Delta$ we show:

Lemma 3.2.1

$S(t)$ is $G(1,\beta)$ extendible (restrictible) onto H_α for all $\alpha \in R$.

Proof

(i) Let $\alpha \in R$. Then we easily see

$$h_k^\alpha := (2|k| + n)^{-\frac{\alpha}{2}} h_k \quad \text{is a CONS in } H_\alpha .$$

Thus, for $\varphi, \psi \in S$

$$<\varphi, \psi>_\alpha = \sum_k <\varphi, h_k^\alpha>_\alpha <\psi, h_k^\alpha>_\alpha$$

$$= \sum_k <\varphi, h_k>_o <\psi, h_k>_o (2|k|+n)^\alpha$$

Hence, for $\varphi, \psi \in S$

$$<\Delta\varphi, \psi>_\alpha = \sum_k <\varphi, \Delta h_k>_o <\psi, \Delta h_k>_o (2|k|+n)^\alpha ,$$

and all we have to show is the existence of a $\beta_\alpha < \infty$ s.t.

$$<\Delta\varphi, \varphi>_\alpha \le \beta_\alpha |\varphi|_\alpha^2 \quad \text{for} \quad \varphi \in S$$

which is equivalent to

$$|S(t)\varphi|_\alpha^2 \le e^{\beta_\alpha t} |\varphi|_\alpha^2 , \quad \varphi \in S .$$

This last relation implies the (unique) extendibility (resp. restrictibility) of $S(t)$ to $S_\alpha(t)$ onto H_α with generator $\frac{1}{2}\Delta_\alpha$ (the extension, resp. restriction, of $\frac{1}{2}\Delta$).

(ii) To simplify notation assume $n = 1$ and note that

$$\partial^2_x g_k(x) = \frac{(k(k-1))^{\frac{1}{2}}}{2} g_{k-2}(x) - \frac{2k+1}{2} g_k(x)$$

$$+ \frac{[(k+1)(k+2)]^{\frac{1}{2}}}{2} g_{k+2}(x)$$

$$x \in R, \quad k \in N \cup \{0\}, \quad g_{-1} \equiv g_{-2} \equiv 0$$

(cf. Holley and Stroock [24], (A8), (A.9)).

Thus, for $\varphi \in S(R)$, setting $\varphi^\alpha_m := <\varphi, g^\alpha_m>_\alpha$, we have $<\Delta\varphi, \varphi>_\alpha$

$$= \Sigma_m \{\varphi^\alpha_m \varphi^\alpha_{m+2} (2m+1)^{\frac{\alpha}{2}} (2m+5)^{-\frac{\alpha}{2}} \frac{[(m+1)(m+2)]^{\frac{1}{2}}}{2}$$

$$- (\varphi^\alpha_m)^2 (\frac{2m+1}{2})$$

$$+ \varphi^\alpha_m \varphi^\alpha_{m-2} (2m+1)^{\frac{\alpha}{2}} (2m-3)^{-\frac{\alpha}{2}} \frac{[(m-1)m]^{\frac{1}{2}}}{2} \}$$

(where we put all terms equal to 0 if the index is smaller than 0)

$$\leq \frac{1}{2} \Sigma_m (\varphi^\alpha_m)^2 \cdot \gamma^\alpha_m$$

with $\gamma^\alpha_m := -(2m+1) + \frac{[m(m-1)]^{\frac{1}{2}}}{2} [(\frac{2m-3}{2m+1})^{\frac{\alpha}{2}} + (\frac{2m+1}{2m-3})^{\frac{\alpha}{2}}]$

$$\frac{[(m+1)(m+2)]^{\frac{1}{2}}}{2} [(\frac{2m+1}{2m+5})^{\frac{\alpha}{2}} + (\frac{2m+5}{2m+1})^{\frac{\alpha}{2}}]$$

and since it is easy to see that

$$\sup_m \gamma_m^\alpha < \infty$$

we may take

$$\beta_\alpha := \frac{1}{2} \sup_m \gamma_m^\alpha .$$

(iii) The n-dimensional case is easily reduced to the onedimensional case by (*) and (3.2.2)

□

Now we can solve the analogue of (3.1.3). Let M be an $H_{-\alpha}$-valued ℓ.s.i. cadlag martingale ($\alpha \geq 0$) and consider for a $\in$ R

$$dY(t) = (\Delta_{-\alpha} + a) y(t) dt + dM(t)$$

$$Y(o) \in H_{-\alpha}$$
(3.2.4)

As a consequence of Lemma 3.2.1 and Th. 2.1.1 (or Remark 2.1.1) we obtain (setting $T_{-\alpha}(t) := S_{-\alpha}(t) e^{at}$):

Corollary 3.2.1

(3.2.4) has a (unique) mild solution Y s.t.

$$Y(\cdot) = T_{-\alpha}(\cdot) Y(0) + \int_0^{\cdot} T_{-\alpha}(\cdot - u) dM(u)$$

$$\in \begin{cases} D([0,\infty); H_{-\alpha}) \text{ a.s. if M is cadlag} \\ C([0,\infty); H_{-\alpha}) \text{ a.s. if M is continous} \end{cases}$$
(3.2.5)

Remark 3.2.2

Since H_α are not generated by Δ alone (but by $(-\Delta + |x|^2)$ -unlike the H_α in Section 3.1) we cannot as in Section 3.1 obtain spatial regularity for the mild solution of (3.2.4) in terms of H_α.

Example 3.2.1

Let a $\in$ R and m_2, V > 0 . Let M be the S'-valued Gaussian martingale with covariance $\int_0^{t \wedge s} <F(u)\varphi, \psi>_0 du$ where $F(u) = -(\nabla e^{at} \nabla) + (m_2 V - a) e^{at}$.

From Itô [27] and Remark 3.2.1 we obtain that M

lives as a continuous process on $H_{-\alpha}$ for all
$\alpha > n + 1$. Hence the mild solution Y to (3.2.4)
lives as a continous Gaussian process on $H_{-\alpha}$ for
all $\alpha > n + 1$ (Cor. 3.2.1). This process was
obtained by Gorostiza [22] as the fluctuation limit
for an infinite system of unscaled branching
Brownian motions. V is the parameter of the
exponential lifetime distribution, m_1 is the
mean of the branching law, m_2 the second
factorial moment and $a := V(m_1-1)$.

Example 3.2.2
If M is the cylindrial Brownian motion on H_o
(i.e., its covariance is $t \wedge s <\varphi,\psi>_o$, $\varphi,\psi \in S$) then
from Itô and Remark 3.2.1 M is an $H_{-\alpha}$-valued
Gaussian martingale for all $\alpha > n$. Then for
$a = o$ the mild solution to (3.2.4) describes the
fluctuation limit obtained by Holley and Stroock
[24] for an infinite system of critial branching
Brownian motions by scaling space and time.

Remark 3.2.3
(i) Another example of (3.2.4) can be found in
    Itô [28].
(ii) Walsh [53] has investigated various scaling
    limits for systems of branching Brownian
    motions on S' and shown that the martingale
    part can be decomposed into two orthogonal
    martingales which account for diffusion and
    branching, respectively (cf. also Kotelenez
    [34] for branching diffusions on a lattice).

Remark 3.2.4
Holley and Stroock [24] have also investigated the
Ornstein-Uhlenbeck process with M(t) cylindrical
Brownian motion and Δ substituted by $\Delta - |x|^2$
in (3.2.4). Clearly, in this case the extension
(restriction) procedure becomes much easier - it is
exactly the same as in Section 3.1. In particular,
the analogue of Lemma 3.1.2 holds, with H_α
substituted by H_α.

3.3 Examples of Second Order Equations on a Bounded Domain

Let $D \subset \mathbb{R}^n$ and A be as in Section 3.1. Moreover, assume

A.3.3.1)
A is negative definite.

Under assumption A.3.3.1) we may take $\beta = 0$ in the construction of the H_α spaces.

Let us consider the unperturbed second order equation:

$$\frac{\partial^2}{\partial t^2} Y(t,r) = AY(t,r) - \mu \frac{\partial}{\partial t} Y(t,r) , \quad \mu \geq 0. \quad (3.3.1)$$

Analogously to the parabolic case we make an extension procedure for the associated first order equation

$$\frac{\partial}{\partial t} X(t,r) = \hat{A} X(t,r)$$

$$X(t) := \begin{pmatrix} Y(t) \\ \frac{\partial}{\partial t} Y(t) \end{pmatrix} , \quad \hat{A} := \begin{pmatrix} O & I \\ A & -\mu \end{pmatrix} \qquad (3.3.2)$$

Note that by Curtain and Pritchard [11]
$$\hat{A} \sim \hat{U}(t) \in G(1,0)$$
on $(\hat{H}_o, (\cdot,\cdot)_o) = (H_1 \times H_o, <\cdot,\cdot>_1 + <\cdot,\cdot>_o)$.

Then, we define for $\alpha \in R$

$$(\hat{H}_\alpha, (\cdot,\cdot)_\alpha) := (H_{\alpha+1} \times H_\alpha, <\cdot,\cdot>_{\alpha+1} + <\cdot,\cdot>_\alpha)$$

and obtain as in Section 3.1 a nuclear Gel'fand triple for (3.3.1):

$$\Phi \subset \hat{H}_\alpha \subset \hat{H}_\gamma \subset \hat{H}_o = \hat{H}_o' \subset \hat{H}_{-\gamma} \subset \hat{H}_{-\alpha} \subset \hat{\Phi}' . \quad (3.3.3)$$

Lemma 3.3.1

$\hat{U}(t)$ is $G(1,0)$ extendible (resp. restrictible) on $\hat{H}_\alpha$ for all $\alpha \in R$.

Proof

(i) One easily verifies $\hat{U}(t)\hat{\Phi} \subset \hat{\Phi}$

(ii) If $\hat{\varphi} \in \hat{\Phi}$ then

$$(\hat{U}(t)\hat{\varphi}, \hat{U}(t)\hat{\varphi})_\alpha = (\hat{\varphi}, \hat{\varphi})_\alpha + \int_0^t 2(\hat{A}\hat{U}(s)\hat{\varphi}, \hat{U}(s)\hat{\varphi})_\alpha ds$$

$$\leq (\hat{\varphi}, \hat{\varphi})_\alpha$$

since for arbitrary $(\varphi, \psi) \in \hat{H}_\alpha = H_{\alpha+1} \times H_\alpha$

$$(\hat{A}(\varphi, \psi), (\varphi, \psi))_\alpha = (\varphi, \psi)_{\alpha+1} + <A\varphi, \psi>_\alpha - \beta|\psi|_\alpha^2$$

$$= -<A\varphi, \psi>_\alpha + <A\varphi, \psi>_\alpha - \beta|\psi|_\alpha = -\beta|\psi|_\alpha^2$$

by the construction of H_α (cf. Section 3.1)

(iii) From this the lemma follows as in Section 3.2.

$\square$

Now let M be an $H_{-\alpha}$-valued martingale and consider the (formal) equation

$$\left. \begin{array}{l} \dfrac{\partial^2}{\partial t^2}Y(t,r) = AY(t,r) - \mu\dfrac{\partial}{\partial t}Y(t,r) + \dfrac{\partial}{\partial t}M(t) \\[2mm] Y(o,r) = Y_o(r), \dfrac{\partial}{\partial t}Y(o,r) = Y_1(r) \end{array} \right\} \quad (3.3.4)$$

which has a meaning as the evolution equation:

$$\left. \begin{array}{l} dX(t) = \hat{A}_{-\alpha}X(t)dt + \begin{pmatrix} 0 \\ dM(t) \end{pmatrix} \\[3mm] X(o) = \begin{pmatrix} Y_o \\ Y_1 \end{pmatrix}. \end{array} \right\} \quad (3.3.5)$$

Corollary 3.3.1

(3.3.5) has a (unique) mild solution

$$\left. \begin{array}{l} X(\cdot) = \hat{U}_{-\alpha}(\cdot)X(o) + \int_0^\cdot \hat{U}_{-\alpha}(\cdot - s)d\begin{pmatrix} 0 \\ M(s) \end{pmatrix} \\[3mm] \in \begin{cases} D([0,\infty); \hat{H}_{-\alpha}) & \text{a.s. if M is cadlag} \\ C([0,\infty); \hat{H}_{-\alpha}) & \text{a.s. if M is continous} \end{cases} \end{array} \right\} \quad (3.3.6)$$

Example 3.3.1

Let $D = [O,L] \subset \mathbb{R}$, $A := \lambda\Delta$, $\lambda > O$,
$M(t) = \sigma W(t)\delta(x-a)$, $\sigma > O$, $a \in [O,L]$, δ the
Delta function, and $W(t)$ one-dimensional standard
Brownian motion. This is an example from continuum
mechanics where a string is exposed to a random
excitation by a point force in a (cf. Wedig [54]).
From Remark 3.1.1 (III) we obtain that $M(t)$ lives
on $\hat{H}_{-\alpha}$ for all $\alpha > \frac{1}{2}$. Hence by Corollary 3.3.1

$X = \begin{pmatrix} Y \\ \frac{\partial}{\partial t}Y \end{pmatrix}$ lives as a continous process on $H_{-\alpha}$,

$\alpha > \frac{1}{2}$ which implies that the solution Y of (3.3.4)

lives on H_ε for all $\varepsilon < \frac{1}{2}$. Moreover, a
criterion of Zabczyk [57] yields the existence of a
unique invariant measure for (3.3.4) under the
preceding assumptions.

Other examples such as those given by Wedig in [55]
can be handled in the same way.

Peter Kotelenez
Forschungsschwerpunkt Dynamische Systeme
Universität Bremen
Bibliothekstraße
Postfach 330 440
2800   Bremen   33
West Germany

4. REFERENCES

[1] S. AGMON: "Lectures on Elliptic Boundary Value
    Problems". Princeton, N. J.: D. van Nostrand
    Co. Inc. (1965)

[2] L. ARNOLD, R. F. CURTAIN, P. KOTELENEZ: "Non-
    linear Evolution Equation in Hilbert Space".
    Forschungsschwerpunkt Dynamische Systeme,
    Universität Bremen, Report Nr. 17 (1980)

[3] A. BAGCHI, V. BORKAR: "Parameter Identification in Infinite Dimensional Linear Systems", Stochastics 12 (3+4) 201-213 (1984)

[4] A. V. BALAKRISHNAN: "On Abstract Stochastic Bilinear Equations with white Noise Inputs", Preprint (1983)

[5] H. T. BANKS, F. KAPPEL: "Spline Approximations for Functional Differential Equations", J. diff. Equations 34 496-522 (1979)

[6] P. BILLINGSLEY: "Convergence of Probability Measures", John Wiley & Sons, New York 1968

[7] P. L. BUTZER, H. BERENS: "Semi-Groups of Operators and Approximation" Springer-Verlag, Berlin-New York 1967

[8] P. R. CHERNOFF: "Two Counterexamples in Semi-group Theory on Hilbert Space", Proc. Amer. Math. Soc. 56 253-255 (1976)

[9] A. CHOJNOWSKA-MICHALIK: "Stochastic Differential Equation in Hilbert Spaces and their Applications", Ph. D. Thesis, Institute of Mathematics, Polish Academy of Science, 1976

[10] R. F. CURTAIN: "Markov Processes Generated by Linear Stochastic Evolution Equations", Stochastic 5, 135-165 (1981)

[11] R. F. CURTAIN, A. J. PRITCHARD: "Infinite Dimensional Linear Systems Theory", LN in Control and Information Sciences 8, Springer-Verlag, Berlin-New York 1978

[12] G. DAPRATO: "Regularity Results of a Convolution Stochastic Integral and Applications to Parabolic Stochastic Equations in a Hilbert Space", Scuola Normale Superiore, Pisa, 1982

[13] G. DAPRATO, P. GRISVARD: "Maximal Regularity for Evolution Equations by Interpolation and Extrapolation" to appear in J. functional Analysis

[14] G. DAPRATO, M. IANNELLI, L. TUBARO: "Some
     Results on Linear Stochastic Differential
     Equations in Hilbert Spaces" Stochastics 6,
     105-116 (1982)

[15] G. DAPRATO, M. IANNELLI, L. TUBARO: "On the Path
     Regularity of a Stochastic Process in a Hilbert
     Space Defined by the Itô Integral", Stochastics
     6, 315-322 (1982)

[16] E. B. DAVIES: "One-Parameter Semigroups",
     Academic Press, London-New York 1980

[17] D. A. DAWSON: "Stochastic Evolution Equations",
     Math. Biosciences 15, 287-316 (1972)

[18] D. A. DAWSON: "Stochastic Evolution Equations
     and Related Measure Processes", J. multivariate
     Analysis 5, 1-52 (1975)

[19] T. FUNAKI: "Random Motion of Strings and
     Related Stochastic Evolution Equations",
     Nagoya math. J. 89, 129-193 (1983)

[20] I. M. GEL'FAND, N. YA. VILENKIN: "Generalized
     Functions", Vol. 4, Academic Press, New York-
     London 1964

[21] I. I. GIHMAN, A. V. SKOROHOD: "The Theory of
     Stochastic Processes", Vol. 1, Springer-Verlag,
     Berlin-New York 1974

[22] L. GOROSTIZA: "High Density Limit Theorems for
     Infinite Systems of Unscaled Branching Brownian
     Motions", Ann. of Probab. 11 (2), 374-392 (1983)

[23] I. GYÖNGY: "On Stochastic Equations with
     Respect to Semimartingales III", Stochastics 7,
     231-254 (1982)

[24] R. HOLLEY, D. W. STROOCK: "Generalized
     Ornstein-Uhlenbeck Processes and Infinite
     Particle Branching Brownian Motions", Publ.
     RIMS, Kyoto Univ. 14, 741-788 (1978)

[25] I. A. IBRAGIMOV: "On Smoothness Conditions for
     Trajectories of Random Functions", Theory
     Probab. Appl., Vol. 28 (2), 240-262 (1983)

[26] A. ICHIKAWA: "Semilinear Stochastic Evolution
     Equations: Boundedness, Stability and Invariant
     Measures", Stochastics 12, 1-39 (1984)

[27] K. ITÔ: "Continuous Additive S' Processes", in
     B. Grigelionis (ed.) "Stochastic Differential
     Systems" Springer-Verlag, Berlin-New York 1980

[28] K ITÔ: "Distribution-Valued Processes Arising
     from Independent Brownian Motions", Math. Z.
     182, 17-33 (1983)

[29] J. JACOD: "Calcul Stochastique et Probléme de
     Martingales", LN in Mathematics 714, Springer-
     Verlag, Berlin-New York 1979

[30] P. KOTELENEZ: "A Submartingale Type Inequality
     with Applications to Stochastic Evolution-
     Equations", Stochastics 8, 139-151 (1982)

[31] P. KOTELENEZ: "Law of Large Numbers and Central
     Limit Theorem for Chemical Reactions with
     Diffusions" Ph. D. Thesis, Bremen 1982

[32] P. KOTELENEZ: "Continuity Properties of Hilbert
     Space Valued Martingales", Stochastic Processes
     Appl. 17 (1), 115-125 (1984)

[33] P. KOTELENEZ: "A Stopped Doob Inequality for
     Stochastic Convolution Integrals and Stochastic
     Evolution Equations" Stoch. Analysis and
     Applications 2 (3), 245-265 (1984)

[34] P. KOTELENEZ: "Linear Parabolic Differential
     Equations as Limits of Space-Time Jump Markov
     Processes" Technical Report Series of the
     Laboratory for Research in Statistics and
     Probability No 38, Carleton University (1984)

[35] N. V. KRYLOV, B. L. ROZOVSKIJ: "On Stochastic
     Evolution Equations", Itogi Nauki i Tehniki,
     Ser. Sov. Probl. Mat. 14 71-146 (1979) (in
     Russian)

[36] H.-H. KUO: "Gaussian Measures in Banach Spaces"
     Springer-Verlag, Berlin-New York 1975

[37] O. A. LADYŽENSKAYA: "The Mixed Problem for a
     Hyperbolic Equation", GITTL, Moscow 1953 (in
     Russian)

[38] J. L. LIONS, E. MAGENES: "Non-Homogeneous Boundary Value Problem I", Springer-Verlag, Berlin-New York 1972

[39] M. METIVIER: "Semimartingales", Walter de Gruyter, Berlin-New York 1982

[40] M. METIVIER, J. PELLAUMAIL: "On a Stopped Doob's Inequality and General Stochastic Equations", Ann. of Probab. $\underline{8}$, Nr. 1, 96-114 (1980)

[41] M. METIVIER, J. PELLAUMAIL: "Stochastic Integration", Academic Press, New York 1980

[42] S.-I. NAKAGIRI: "On the Fundamental Solution of Delay-Differential Equations in Banach Spaces", J. diff. Equations 41, 349-368 (1981)

[43] E. PARDOUX: "Equations aux derivees partielles stochastic non linéaires monotones. Etude de solutions fortes de type Itô", These doct. Sci. math. Univ. Paris Sud, 1975

[44] E. PARDOUX: "Stochastic Partial Differential Equations and Filtering of Diffusion Processes", Stochastics $\underline{3}$, 127-167 (1979)

[45] B. L. ROZOVSKIJ: "Stochastic Evolution Systems", Nauka, Moscow 1983 (in Russian)

[46] H. H. SCHAEFER: "Topological Vector Spaces", Springer-Verlag, New York-Berlin 1980

[47] H. TANABE: "Equations of Evolution", Pitman, London-San-Francisco-Melbourne 1979

[48] H. TRIEBEL: "Interpolation Theory, Function Spaces, Differential Operators", North-Holland, Amsterdam-New York-Oxford 1978

[49] L. TUBARO: "An Estimate of Burkholder Type for Stochastic Processes Defined by the Stochastic Integral", Stoch. Analysis and Applications, 187-192 (1984)

[50] C. TUDOR: "Some Properties of Mild Solutions of Delay Stochastic Evolution Equations", Preprint (1984)

[51] A. S. USTUNEL: "Additive Processes on Nuclear Spaces" Ann. of Probab. 12, Nr. 3, 858-868 (1984)

[52] J. WALSH: "A Stochastic Model of Neural Response" Adv. Appl. Prob. 13, 231-281 (1981)

[53] J. WALSH: Preprint (1984)

[54] W. WEDIG: "Covariance Analysis of Distributed Systems under Stochastic Point Forces" Preprint 1982

[55] W. WEDIG: "Moments and Probability Densities of Parametrically Excited Systems and Continuous Systems" Abhandlungen der Adademic der Wissenschaften der DDR Abt. Mathematik-Naturwissenschaften-Technik 6N (1977)

[56] T. A. WITTIG: "A Markov Characterization of Generalized Evolution Processes", Preprint

[57] J. ZABCZYK: "Linear Stochastic Systems in Hilbert Space; Structural Properties and Limit Behaviour", Institute of Mathematics, Polish Academy of Sciences, March 1981, Preprint No 236

Paul Krée

MARKOVIANIZATION OF RANDOM VIBRATIONS

1. INTRODUCTION

The infinite dimensional filtering theory is today well
developped and important theoretical and practical problems
in this aera have been solved : see for example the present
volume and [4], [6], [11] .
 This lecture is motivated by the constructive study of
non linear random vibrations governed by stochastic differential equations (SDE)

$$\dot{\eta}_t = f(\eta_t , exc_t) \qquad \text{for } t > 0 \qquad\qquad (1.1)$$

with a given initial condition for $t = 0$. The driving
process (exc_t) called the excitation is usually colored.
The uniqueness and the existence for all $t > 0$ of the
solution is assumed : see [10] p.10-14. Currently (exc_t) is
a physically realizable and stationary Gaussian process with
an irrational spectral density

$$S(\omega) = (2\pi)^{-1} \mathscr{F} C(\omega) = (2\pi)^{-1} \int_{-\infty}^{+\infty} C(t) e^{-it\omega} dt \qquad (1.2)$$

where C denotes the correlation function In aeronautics
for example, exc_t is the vertical component V_z of the
turbulent wind V^t and has for spectral density :

$$S_V(\omega) = (a^2 + b^2 \omega^2)(c^2 + d^2 \omega^2)^{-11/6} \qquad (1.3)$$

with $a, b, c,$ and $d > 0$. In civil engineering the horizontal component V_x of V in the direction 0_x of
dominant winds is modelized in any plane $P \perp 0_x$ by a
stationary Gaussian process with interspectral measures.

$$S(m,m',\omega) = \omega(1 + k\omega^2)^{-4/3} e^{-k'\omega(\alpha\Delta y^2 + \beta\Delta z^2)^{1/2}} \qquad (1.4)$$

with $m = (y,z)$ and $m' = (y',z') \in P$, $\Delta y = y-y'$,
$\Delta z = z-z'$, $k , k' , \alpha , \beta > 0$.

141

L. Arnold and P. Kotelenez (eds.), Stochastic Space-Time Models and Limit Theorems, 141–162.
© *1985 by D. Reidel Publishing Company.*

Also for structures submitted to the action of waves, the
components of exc_t are built up by integration of some
Gaussian random fields on various parts of the structure em-
bedded in the random field : see [13] chapters VI and VIII.
     Constructive methods for the study of random vibration
use the Markoff property, the Fokker-Planck Equation (PE)...,
hence are only valid for Ito's stochastic equations where the
driving process is a Gaussian white noise $N_t = \dot{W}_t = dW_t / dt$
     Hence these methods are not valid for(1.1). But
suppose that a linear and time independent differential
filter driven by a white noise

$$\dot{\xi}_t = A \xi_t + Q \dot{W}_t \qquad (1.5)$$

and admitting a stationary solution (ξ_t) can be found such
that exc_t is a linear observation of ξ_t where (ξ_t) has no
infinite past

$$exc_t = B \xi_t \qquad (1.6)$$

where B is linear. Then the elimination of exc_t between
(I.1) and (I.6) would give an Ito's equation for the augmented process $\zeta_t = (\xi_t , \eta_t)$:

$$\begin{aligned} \dot{\xi}_t &= A \xi_t + Q \dot{W}_t \\ \dot{\eta}_t &= f(t , B \xi_t) \end{aligned} \quad\Longleftrightarrow\quad \begin{aligned} d\xi &= A \xi_t \, dt + Q \, dW \\ d\eta &= f(t , B \xi_t) \, dt \end{aligned} \qquad (1.7)$$

Hence the constructive methods can be applied to (I.7). We
say that (1.1) has been Markovianized since the augmented
process (ζ_t) has the Markoff property. Hence the following
filtering problems arise in the study of random vibrations :
(I.8) Problem 1 : For a given stationary and Gaussian physically realizable process exc_t with an irrational spectrum,
find a linear differential filter (I.5) admitting a stationary state (ξ_t) such that (1.6) holds for a linear mapping
B. What is the class of these linear filters ? This class is
well known in the finite dimensional case since then the
matrix A is asymptotically stable : $\lambda \in$ Sp $A \Rightarrow$ Re $\lambda < 0$.
Note that ξ cannot be a finite dimensional process since
then $B\xi$ would necessarily have a rational spectral density.
Since computers work only with a finite information, the
solution of problem 1 is not sufficient and engineers are
interested by approximate Markovianization :
(1.9) Problem 2 : Make the Markovianization sufficiently
explicit in order to help engineers in the construction of
finite dimensional approximations of the infinite dimensional
filter (1.5)

(1.10) <u>Problems 3,4</u> ... Extend the mathematical theories
concerning usual Stochastic Differential Equation (SDE)
such as P. Malliavin's theory [20] [23] to the infinite
dimensional SDE of type (1.7). This needs for example an
extension of the usual theory of Sobolev spaces on Gaussian
spaces [19] [17] [18] ... to Banach-valued Sobolev spaces
with respect to non Gaussian measures.
     My contribution to the Bremen workshop was mainly con-
cerned with the solution published in [14] of problems 1 and
2. This solution is presented briefly and in a different way
in parts II, III, IV. In particular further examples and references showing the mathematical interest of asymptotically
stable linear systems and the practical interest of the approximate Markovianization are given in parts III and IV resp.
In order to give more interest to this written version of the
lecture, some preliminary results concerning problems 3,4 ...
are given in parts V and VI. These new results are obtained
using the absolute and the relative differential calculus
for measures developed some time ago [17] [18] [19] and
applied at that time only to Gaussian measures and to Hilbert-
valued functions. Three theorems are proved. The first is of
a new kind. The last two theorems reinforce two theorems
of P. Malliavin's theory in the following sense : using
weaker hypothesis stronger conclusions are proved. Note that
Sobolev spaces $W^{p,1}$ <u>of order one</u> with resp to a quasi-invariant and non Gaussian measure M have been studied previously [1] [20] [26]. In the present case M is not necessarely quasi - invariant and all Sobolev spaces $W^{p,k}$ will
be defined and used.

2. CYLINDRICAL FORMALISM IN THE LINEAR FILTERING THEORY

The language and the concepts of the elementary linear
filtering theory (see for ex [9] chap. V and [13] chap. IV)
are weakened in order to obtain a formulation that is also
valid in the case of infinite dimensional filters. But there
is no miracle : in each given situation the weak results
obtained in this way have to be sharpened in a second step
in order to eliminate the cylindrical concepts.
     This weakening is realized below in the following way :

- a finite dimensional vector space V is viewed as the
space $(V')'$ of all linear forms on the dual space V'

- any V-valued random variable ξ is viewed as a cylindrical V-valued random variable i.e. as a linear mapping
(linear process)

$$V' \ni u \longrightarrow u \circ \xi = < \xi , u > \in L^{o}(\Omega) \tag{2.1}$$

- For any vector space W , a W- valued linear observation $B \xi$ of ξ is identified with the corresponding cylindrical random vector of W i.e. with the product of (2.1) with the transpose B^T of B . Hence :

$$\forall \omega' \in W' \quad < B \xi , \omega' > = < \xi , B^T \omega' > \tag{2.2}$$

. matrices are identified with the corresponding bilinear forms. For example the familiar definition of the covariance of a V-valued stationary process $q = (q_s)$ on the line

$$C_q(s-t) = E [q_s q_t^T] \tag{2.3}$$

with $V \simeq \mathbb{R}^d$ is replaced by the following where $C_q(s-t)$ is now viewed as a bilinear form on $V' \times V'$

$$V \times V' \ni (v,v') \longrightarrow < C_q(s-t)v,v' > = E [< q_s,v' > < q_t,v >] \tag{2.4}$$

. Hence for any locally convex Hausdorff space (ℓ.c.H.s) Y a cylindrical random vector of Y' is defined by an element of the space $L(Y,L^2(\Omega))$ of all linear and continuous mappings $Y \to L^2(\Omega)$. Endowing this space with the topology of simple convergence a Y'-valued cylindrical process on the line is defined by a continuous mapping :

$$\mathbb{R} \ni t \longrightarrow \vec{\xi}_t \in L(Y,L^2(\Omega)) \tag{2.5}$$

White noises are defined by generalized cylindrical processes i.e. by linear and continuous mappings

$$\mathcal{D}(\mathbb{R}) = \mathcal{B} \ni \varphi \to \vec{\xi}_\varphi \in L(Y,L^2(\Omega)) \tag{2.6}$$

This can be equivalently defined by linear mappings :

$$\mathcal{D} \otimes Y \ni \varphi \otimes y \longrightarrow < \xi_\varphi , y > \in L^2(\Omega) \tag{2.7}$$

separately continuous in theis arguments φ and y

(2.8) <u>Gaussian white noise N with co variance</u>

$$C_N(s-t) = Q \delta_o(s-t) \tag{2.9}$$

<u>where $Q \in L_+(Y,Y')$ is given.</u> Let j be the canonical surjection $Y \to Y/\ker Q$. Let j_Q be the product of j with

the canonical injection of $Y/\mathrm{Ker}\ Q$ with its completion for the scalar product defined by Q. Then (Y_Q, j_Q) is called the Hilbert sursurface of X defined by Q. Identifying isometrically $X_N = L^2(\mathbb{R}, X_Q)$ with a space of (centered) Gaussian random variables, N is defined as the product of the following two linear mappings

$$\mathcal{D} \otimes Y \xrightarrow{\quad j \otimes j_Q \quad} L^2(\mathbb{R}) \otimes Y_Q \xrightarrow{\quad\quad} Y_N \qquad (2.10)$$

where j denotes the injection of $\mathcal{D}$ in $L^2(\mathbb{R})$

(2.11) Weak formulation of the elementary linear filtering theory. The familiar definition of the filtering of an $\mathbb{R}^d$-valued white noise N by a causal filter $k \in L^2(\mathbb{R}^+, \mathrm{Mat}(d', d))$:

$$N \longrightarrow (k * N)_t = \int_{-\infty}^{t} k(t-s)\ N_s\ ds \qquad (2.12)$$

has non mathematical meaning ([9] p.284) and [8] gives no meaning to the last integral since $k(t-.) \notin \mathcal{D}$. Following [13], chap.12 a meaning is given to the last integral defining the cylindrical vector associated with the random variable $(k * N)_t$ by

$$(\mathbb{R}^{d'})^* \ni u \rightarrow \langle k * N)_t, u \rangle = \int N_s\ k(t-s)^T u\ ds$$

Hence

$$\langle (k * N)_t, u \rangle = \langle N_., k(t-.)^T u \rangle \qquad (2.13)$$

= random variable associated by N with the element $k(t-.)^T u \in L^2(\mathbb{R}, \mathbb{R}^d)$

The formula (2.13) gives a weak dimension free-formulation of the convolution filtering. In fact let X be a $\ell.c.H.s$, a white noise N of type (2.8) and an X-valued causal filter k adapted to N i.e. a linear and continuous mapping

$$X \ni x \longrightarrow k(.)^T x \in L^2(\mathbb{R}^+, Y) \qquad (2.14)$$

Then $k * N$ is a process on the line with values cylindrical vectors of X', and $k * N$ is defined by (2.13).

Linear filtering is defined in the same way. For example the Wiener-process W of covariance $Q \min(s,t)$ on

$\mathbb{R}^+$ x $\mathbb{R}^+$ is defined by integration on $\mathbb{R}^+$ of the white noice (2.8). Therefore $\dot{W} = N$ and the right hand side of (2.13) can also be viewed as a Wiener integral [12]

(2.15) Finally the definition of the Reproducing Hilbert space H_Q of a positive quadratic form $<Qy,y>$ on some $\ell.c.H.s.$ Y is recalled : H_Q is the completion of the image of $Q \in L_+(Y,Y')$ for the following scalar product on Im Q :

$$[Qy , Qy'] = < Qy , y' >$$ (2.16)

(2.17) In general the reproducing space H_Q and the Hilbertian surspace Y_Q of a positive quadratic form $< Qy,y >$ on Y are in duality : see [13] § 10.7.

For example if ξ is a centered cylindrical random vector of Y' , the covariance of ξ is the following quadratic form

$$< Q y , y' > = E [< \xi , y > < \xi , y' >]$$

and H_Q is called the reproducing Hilbert space of ξ : see [13] § XI.6.

3. STATIONARY GAUSSIAN AND MARKOVIAN CYLINDRICAL PROCESSES

Let X be a $\ell.c.H.s.$ Let (ξ_t) be a stationnary X'-valued cylindrical Gaussian process with injective covariance C . Below X is Hilbertian since the space X is replaced by its completion for the scalar product $< Cx,x' >$. For any s, M_s^-(resp M_s) denotes the closed linear space of $L^2(\Omega)$ generated by $\{ < \xi_t , x > , t \leqslant s$ and $x \in X \}$ (resp of $\{ < \xi_s > , x \in X \}$. Denoting $Proj_s$ and $\overline{Proj}_s$ the operators of orthogonal projection in $L^2(\Omega)$ on M_s and M_s^- resp (ξ_t) is called weakly Markovian if

$$x \in X \quad and \quad s \leqslant t \Rightarrow \overline{Proj}_s (< \xi_t,x >) = Proj_s(< \xi_t,x >)$$

(3.2) Lemma. For arbitrary $t > 0$ and $x \in X$, let $S_t x$ be the element of X such that for all reals s

$$(3.3) \quad Proj_s(< \xi_{s+t} , x >) = < \xi_s , S_t x >$$

Then S_t is a linear contraction in X

(3.4) $\forall \ x \in X \quad \| \ S_t \ x - x \ \| \rightarrow 0 \quad$ if $\quad t \downarrow 0$

and the covariance of ξ is :

$$E \left[< \xi_s \ , \ x > < \xi_t \ , \ y > \right] = \begin{cases} <x \ , \ S_{t-s} \ y > \quad \text{for } t \geqslant s \\ <S_{s-t} \ x \ , \ y> \quad \text{for } s \geqslant t \end{cases} \qquad (3.5)$$

In fact $\ S_t \ $ is contracting for all $\ t \geqslant 0 \ $ since

$$\| \ S_t \ x \ \| = \| < \xi_0 \ , S_t x > \| = \| \text{Proj}_0 < \xi_t , x > \| \leqslant \| < \xi_t , x > \| = \| \ x \ \|$$

and for $\ t \downarrow 0$

$$\| \ S_t \ x - x \ \| = \| < \xi_0 \ , \ S_t x > - < \xi_0 \ , \ x > \|$$

$$= \| \text{Proj}_0 (< \xi_t \ , \ x > - < \xi_0 \ , \ x >) \| \leqslant \| < \xi_t \ , \ x> - < \xi_0 , \ x> \| \rightarrow 0$$

(3.6) <u>Proposition</u>. The X'valued stationary cylindrical
Gaussian process ξ is weakly Markovian iff
$\{S_t \ , \ t \geqslant 0\}$ is a semi-group $\in L(X)$

For the necessity [12] we have denoting $\text{Proj}_{0,s}$ the orthogonal projection on the Gaussian subspace generated by $\ \xi_0$
and ξ_s in $L^2(\Omega)$

$$< \xi_0 \ , \ S_{s+t} \ x> = \text{Proj}_0 \ < \xi_{s+t} , \ x> = \text{Proj}_0 (\text{Proj}_{0,s} < \xi_{s+t} , x>)$$

$$= \text{Proj}_0 (\text{Proj}_s < \xi_{s+t} , \ x>) = \text{Proj}_0 < \xi_s , \ S_t \ x> = < \xi_0 , \ S_s \ S_t \ x>$$

Hence (S_t) is a semi-group. Conversely it (S_t) is a
semi-group.

$$\text{Proj}_0 \ \text{Proj}_{0,s} \ < \xi_{s+t} , \ x> = \text{Proj}_0 (\text{Proj}_s \ < \xi_{s+t} , \ x>)$$

Hence $\ \text{Proj}_{0,s} \ < \xi_{s+t} , \ x>= \text{Proj}_s \ < \xi_{s+t} , \ x>$

By induction for $\ s_1 < s_2 < \ldots < s_n < t$

$$\text{Proj}_{s_n} < \xi_t , \ x> = \text{Proj}_{s_1,s_2,\ldots,s_n} < \xi_t , \ x>$$

Hence ξ is weakly Markovian and the Proposition is proved

If $\ \xi$ is weakly Markovian the domain D of the generator
A of the semi-group (S_t) is endoved with the graph norm,
the canonical injection $\ i \ : D \rightarrow X$ is continuous. Denoting

A^t the transpose of the linear operator A of X , the transpose of $A \in L(D,X)$ is denoted $A^T_{ex} \in L(X',D')$ and extends A^T

$$D \xrightarrow[\underset{\scriptstyle i}{\longleftrightarrow}]{A} X \simeq X' \xrightarrow[\underset{\scriptstyle i^T}{\longleftrightarrow}]{A^T_{ex}} D' \tag{3.7}$$

(3.8) <u>Proposition</u>. Let ξ be an X'-valued Gaussian and weakly Markovian cylindrical process. Then the D'-valued cylindrical process

$$<W_t, x> = <\xi_t, x> - <\xi_o, x> - \int_o^t <\xi_u, A_x> du \tag{3.9}$$

is a Wiener cylindrical process with covariance Q min (s,t) on $\mathbb{R}^+$ where

$$Q = i^T \circ A - A^T_{ex} \circ i \tag{3.10}$$

In fact (3.9) defines W as a Gaussian (and centered) cylindrical process. Hence (3.8) is proved simply computing the covariance of W

(3.11) <u>Corollary</u>. The derivation with respect to t of generalized cylindrical processes gives starting from (3.9) :

$$\dot{\xi} = - A^T \xi + \dot{W} \tag{3.12}$$

where the D'-valued cylindrical process $A^T \xi$ is defined by $<A^T \xi, \varphi> = <\xi, A\varphi>$.

Conversely we start with an asymptotically stable linear system i.e. with an Hilbert space X and a continuous semi-group (exp t A , t > 0) of contractions in L(X) satisfying the following condition

$$\forall x \in X \ \| (\exp t A) x \| \to 0 \quad if \quad t \to \infty \tag{C}$$

Defining Q by (3.10) we are interested by the existence and uniqueness of a D'-valued Gaussian and weakly Markovian cylindrical process ξ solution of (3.12). Introducing also the Hilbert surspace (D_Q, j_Q) of Q we have for arbitrary $x \in D$ [21][14]

$$\int_o^\infty \| j_Q e^{tA} x \|^2 \ dt = \lim_{T \to \infty} \int_0^T ... = \lim (\|x\|^2 - \|e^{tA} x\|^2 = \| x \|^2$$

(3.13) Hence a linear and isometric imbedding

$$X \ni x \xrightarrow{\ k'\ } (t \rightarrow j_Q \ e_+^{tA} \ x) \in L^2(\mathbb{R}, D_Q)$$

(3.14) <u>Theorem</u> [12][14][21]. Let X be an Hilbert space
and let {exp tA, t > 0} be a continuous semi-group
of contractions $\in L(X)$ satisfying (C)
a) Let N = W be a Gaussian white noise with covariance
Q δ_o(s-t) where Q is given by

(3.10) Then the D'-valued cylindrical process $\xi = k'*N$
is stationary, Gaussian weakly Markovian and satisfies
(3.12). Conversely Let ξ be a D'-valued cylindrical
stationary Gaussian process ξ without infinite past.
i.e. $\cap M_t = \{0\}$ where M_t denotes the closed subspace
of $L^2(\Omega)$ generated by the variables $<\xi_s, x>$, s < t
and $x \in X$. Then $\xi = k'*N$.

(3.15) <u>The two Hilbert spaces defined by a weak asymptotically stable linear filter</u>. Denoting ξ the stationary solution
and C δ_o(s-t) the covariance of the corresponding white
noise, these two spaces are respectively the reproducing
Hilbert space H_ξ of ξ_t and the Hilbert surspace
(j_Q, D_Q) of Q .

    Since D_Q is the Gaussian space of random variables dri-
ving ξ , D_Q is appropriate for the stochastic calculus of
variations [22] concerning gentle perturbations of ξ .
And since m = Law ξ_t has a good behaviour with resp. to all
translations in H_ξ , H_ξ is appropriate for the formulation
of the weak differential calculus relatively to m : see part
VI.

    The particular case where $D_Q = H_\xi$ is the simplest

(3.16) <u>The P. Malliavin's linear filter</u> [22] is defined by an
infinite dimensional and separable real Hilbert space X
and by the continuous semi-group {Id exp - t/2 , t > 0} in
L(X) , where Id denotes the identity mapping of X . In
fact, in this case D = X = D' and Q = Id . Hence the two
Hilbert spaces H_ξ and D_Q coincide. The solutions of the
corresponding stochastic equation

$$\dot{\xi} + \xi/2 = \dot{W} = N \Longleftrightarrow d\xi = -\xi dt/2 + dW \qquad (3.17)$$

define a Markoff process living in some completion $\tilde{X}$ of X

and the canonical cylindrical measure of X is invariant.

In order to build up a process of the same nature but living in X , the drift term $- \xi \, dt$ in (3.17) can be replaced by a stronger one :

(3.18) <u>The B. Gaveau's linear filter</u>. Let u be a symmetric injective Hilbert Schmidt operator of a separable Hilbert space X . Let (e_n) be an orthonormal basis of X .

$$u(e_n) = \lambda_n^{-1/2} \, e_n \quad , \quad \Sigma \, \lambda_n^{-1} < \infty$$

Hence $X \sim \ell_2 = \{x = \Sigma \, x_n' \, e_n \; ; \; \Sigma \, x_n'^2 < \infty\}$

In [7] the following X-valued stochastic equation is defined

$$d \, \xi_n = - \, \lambda_n \, \xi_n/2 + d \, W_n \qquad n = 1,2 \qquad\qquad (3.19)$$

where the Wiener processes W_n are independent

The process $\xi = \xi_n)$ admits the invariant measure

$$\gamma = \prod_{n=1}^{\infty} \gamma_n \quad \text{with} \quad \gamma_n = \left(\frac{\lambda_n}{2\pi}\right)^{1/2} \left(\exp - \frac{\lambda_n}{2} x_n'^2\right) dx_n'$$

Let H_R be the Hilbert space admitting the $\varepsilon_n = \lambda_n^{-1/2} \, e_n$ as an orthonormal basis :

$$H_R = \{ \Sigma_1^{\infty} x_n \, \varepsilon_n \; ; \; \Sigma \, x_n^2 < \infty\}$$

Putting

$$x_n' = \lambda_n^{-1/2} \, x_n \qquad\qquad (3.20)$$

the canonical injection of H_R in X :

$$H_R \ni (x_n) \xrightarrow{\;i\;} x' = (x_n') \in X$$

is Hilbert-Schmidt and maps the canonical cylindrical measure of H_R on the Radon measure γ . This means that H_R is the reproducing Hilbert space of γ . The Gaussian Hilbert space defined by the driving process (W_n) is $X \neq H_R$. Therefore the coordinate transformation (3.20) permits the conversion of the usual formulation (with x_n coordinates) of results concerning the differential calculus connected with γ to a formulation compatible with the stochastic calculus of variation (using x_n' coordinates).

4. APPLICATION TO MARKOVIANIZATION [14]

The method is explained below. Let us consider a random
vibration driven by a scalar physically realizable process

$$exc_t = k * N \tag{4.1}$$

with $k \in X = L^2(\mathbb{R}^+)$ and where N denotes a stationary
scalar Gaussian white noise with covariance $\delta_o(s-t)$. For
example $exc_t = V$ with spectral measure (1.3), the Laplace
transform of k is

$$p \rightarrow \int_o^\infty e^{-pt} k(t) \, dt = \frac{a + bp}{(c + dp)^{11/6}}$$

The following semi-group in $L(X)$

$$\exp tA : f \rightarrow \tau_t f = f(t + .) \; ; \quad t > 0 \tag{4.2}$$

is a continuous semi-group of contractions in $L(X)$ satisfying the condition (C) .

(4.3) The corresponding weak asymptotically stable linear
filter is constructed using the procedure (3.7). Hence the
domain D of $A = d/dt$ in the Sobolev space $H^1(\mathbb{R})$. The
dual D' is the subspace $\overset{o-1}{H}(\mathbb{R}^+)$ of elements $u \in H^{-1}(\mathbb{R})$
supported by $[0 , + \infty[$. Hence Q is $\varphi \rightarrow \varphi(0) \delta_o$.

Therefore D_Q is $\mathbb{R}$ and j_Q is

$$D_Q \ni \varphi \xrightarrow{\ j_Q\ } \varphi(0) \in \mathbb{R} \tag{4.4}$$

The dynamic equation of this filter is (1.5) where
$A^T = - d/dt$ and where N has covariance $Q \delta_o(s-t)$

(4.5) Proposition [14]. For any causal filter $k \in X$, the
physically realizable process $(k * N)_t$ coincides with
the linear observation $B\xi_t$ of the stationary state ξ
of (I.5) where B denotes the linear form on X defined by k :

$$B \xi_t = < \xi_t , k > \tag{4.6}$$

The brackets have a meaning since k belongs to the Reproducing Hilbert space of ξ_t . Note that the complete definition of the Markovianization $\zeta = (\xi, \eta)$ cf (1.1) needs
the defintion of all probalities of transition, hence the

introduction of some completion $\hat{X}$ of X where the random
vectors ξ_t take their values.

Proof of (4.5). According to the definition (3.12) of ξ
by filtering N :

$$B \, \xi_t = \langle \xi_t \, , \, k \rangle = \langle N_u \, , \, j_Q \, e^{(t-u)A} \, k \rangle$$

Hence using (4.2)

$$= \langle N_u \, , \, j_Q(k(t-u + .)) \rangle = \langle N_u \, , \, k(t-u) \rangle = (k * N)_t$$

Note that the driving white noise N is identical with the
white noise expressing exc_t in (4.1). Hence the Markovianization does not need an enlarging of the filtration

(4.7) Reduction of the Markovianization :
In the previous construction , $X = L^2(\mathbb{R}^+)$ can be replaced
by the closed subspace X_h of X generated by all right
translated functions $h(t + .)$ of h . For example, if
$N = dim \, X_h < \infty$, the vibration driven by h can be Markovianized by simply adding N components to ξ .

The Markovianization of vibrations driven by a vectorial
process or by a physically realizable random field uses the
same method : see [14].

(4.8) Approximate Markovianization :
Denoting L_ℓ the ℓth Laguerre polynomial and putting

$$e_\ell = (\ell!)^{-1} L_\ell(x) \, exp(-x/2)$$

then $\{e_\ell , \ell = 0,1 \dots\}$ is an orthonormal basis of X such
that for any M , the subspace X_M of X generated by
$e_o, e_1, \dots e_M$ is invariant by all left translation. Hence in
view of (4.7) an approximate Markovianization of (1.1) driven
by $exc_t = k * N$ is constructed proceeding as follows :

. The element $k \in X$ is expanded in term's of the e_ℓ :

$$k \sim \Sigma_o^M k_\ell e_\ell \qquad\qquad k_\ell = \langle k \, , \, e_\ell \rangle$$

. The auxiliary dynamic system is defined by the restriction
of $\{exp \, A \, t\}$ to the subspace X_M of X

(4.9) Some recent progress concerning approximate Markovianization.
References [2] [3] give applications for simulation procedu-

res. Reference [27] refers to numerical algorithms concerning
fatigue loads of off shore structures.

5. THE WEAK ABSOLUTE DIFFERENTIAL CALCULUS FOR MEASURES.

Let X be a ℓ.c.H.s. The family $\{X', \alpha \in A\}$ of all finite
dimensional subspaces of the dual space X' is directed for
the ordering defined by the inclusion relation. The notation
$X'_\alpha \subseteq X'$ means that X'_α is a finite dimensional subspace
of X' . Hence an inductive system $\{X'_\alpha , i_{\alpha\beta}\}$ of finite
dimensional vector spaces. Hence transposing and putting X_α
$= (X'_\alpha)' \simeq X/(X'^{\perp}_\alpha)$ an inductive system $\{X_\alpha , s_{\alpha\beta}\}$ of vector
spaces and a lifting (X , s_α) of this system are constructed

Hence a diagram

$$
\begin{array}{ccccc}
 & \xrightarrow{\quad f \quad} & & V & \\
 & & & \uparrow f_\beta & \\
X & \xrightarrow{s_\alpha} & X_\alpha & \xrightarrow{s_{\alpha\beta}} & X_\beta \\
\vdots & & \vdots & & \vdots \\
X' & \xleftarrow{i_\alpha} & X'_\alpha & \xleftarrow{i_{\alpha\beta}} & X'_\beta
\end{array}
\qquad (5.1)
$$

where doted lines represent dualities. For any ℓ.c.H.s.
V a V-valued function f on X is called cylindrical if
$f = s_\beta \circ f_\beta$ for some $\beta \in A$ and some $f_\beta : X_\beta \to V$ and X_β
is called a basis of f . The space $C_o^\infty(X_\alpha)$ of C^α functions
φ on X_α such that for any fixed j , $\nabla^j \varphi(x) \to 0$ if
$|x| \to \infty$ has a natural Frechet topology. The Hahn Banach
theorem shows that the dual of $\mathcal{C}(X_\alpha) = 0_c(X_\alpha)$
$= \lim_{\to k} (1 + |x|^2)^k C_o^\infty(X_\alpha)$ is the L. Schwartz $0'_c(X_\alpha)$ of
all convolution operators in $\mathcal{S}(X_\alpha)$. For $\beta < \alpha$ the mapping

$$
\mathcal{C}(X_\beta) \ni F \longrightarrow F \circ s_{\alpha\beta} \in \mathcal{C}(X_\alpha) \qquad (5.2)
$$

identifies $\mathcal{C}(X_\beta)$ to a topological subspace of $\mathcal{C}(X_\alpha)$.
A ℓ.c.H.s. of cylindrical test functions on X is defined
putting

$$
\mathcal{C}_{cyl}(X) = \lim_{\to} \mathcal{C}(X_\alpha) \qquad (5.3)
$$

Following [17] [19], an element T of the space
$L(\mathcal{C}_{cyl}(X) , V)$ is called a V-valued cylindrical distribution on X . Hence T is characterized by a coherent family
of V-valued distributions T_α on the spaces X_α .

(5.4) Let M denote an interiorely regular probability measure on the Baire σ-field of X s.t. $\mathcal{C}_{cyl}(X) \subset \cap L_M^p(X)$. Hence M is characterized by the cylindrical distribution (M_α) on X with $M_\alpha = (s_\alpha)_*(M)$. We also assume that the mappings s_α ($\alpha \in A$) generate the completed σ-field of X.

(5.5) <u>Martingale representation of any $F \in L_M^p(X,V)$ for $1 < p < \infty$ and for a Banach space V with the Radon - Nikodym property</u>. In fact, F is characterized by the cylindrical distribution $(F_\alpha M_\alpha)$ where the L^p-norms of the F_α $= E[F \| s_\alpha]$ are uniformly bounded [17] [19].

(5.6) <u>Weak derivative $\nabla^k M$ for any given $M \in L(\mathcal{C}_{cyl}(X),V)$</u>. For any integer k > 0 a multilinear and symmetric mapping :

$$X^k \ni (x_1,\ldots,x_n) \to \partial_{x_1 x_2 \ldots x_k} M \in L(\mathcal{C}_{cyl}(X),V) \tag{5.7}$$

is defined recursively putting

$$< \partial_x M , F > = - < M , \partial_x F > \tag{5.8}$$

As in [17] [19] the global weak derivative $\nabla^k M$ of order k is defined by the linear mapping

$$\odot_k X \xrightarrow{\nabla^k M} L(\mathcal{C}_{cyl}(X) , V) \tag{5.9}$$

defined by (5.7) or equivalently by the corresponding linear mapping

$$\mathcal{C}_{cyl}(X) \xrightarrow{\nabla^k M} Mul_{sym}(X^k , V) \tag{5.10}$$

Hence $\nabla^k M$ is characterized by the coherent family of derivatives $\nabla^k M_\alpha : \mathcal{C}(X_\alpha) \to Mult_{sym}(X_k,V)$ of the measures M_α

(5.11) <u>H-smoothness of probability measures on X</u>

Let H be a ℓ.c.H.s. continuously and linearly injected in X . A probability M on X is called k-times H-derivable if M satisfies (5.4) and if for arbitrary j < k and 1 < p ≤ ∞ , the linear mapping $\mathcal{C}_{cyl}(X) \to Mul_{sym}(H^j,\mathbb{R})$ admits a continuous extension :

$$\nabla^j M \in L(L_M^p(X) , Mul_{sym}(H^j,\mathbb{R})) \tag{5.12}$$

Usually the vectorial measure $\nabla^j M$ is absolutely continuous with respect to M i.e.

$$\nabla^j M = C_j M \tag{5.13}$$

where the function C_j takes values in some completion of $\text{Mul}_{\text{sym}}(H^j, \mathbb{R})$. In the same way :

$$\nabla^j M_\alpha = C_{j,\alpha} M_\alpha \text{ with } M_\alpha \in L^{p'}_{M(\alpha)}(X, \odot_j X_\alpha) \tag{5.14}$$

Hence for arbitrary $h \in H$

$$\nabla^j M \cdot h^j = (\partial_h)^j M = C_j(x) . h^j M (dx) \tag{5.15}$$

(5.16) Remarks. a) If M is k-times H.derivable for $k = 1,2 \ldots$ then M is not necessarely H-quasi-invariant : see for example on $X = H = \mathbb{R}$ the C^∞ measure $M(x)$ vanishing for $x < 0$ and $= \exp{-x^{-1}}$ for $0 < x < 1$
b) But if M is H-quasi-invariant and $H-C^1$, our definition of ∇M is coherent with the usual one using translations. In fact for arbitrary $h \in H$ and $\lambda \to 0$

$$\partial_h M = \lim \lambda^{-1}(\tau_{\lambda h} M - M) \text{ in } \mathcal{C}_{\text{cyl}}(X)' \text{ weak}$$

(5.17) Theorem 1. With the previous notations let M be an $H-C^\infty$ probability measure one X such that for arbitrary $p \in]1, + \infty[$

$$\sum_{\ell=o}^{\infty} \ell!^{-1} \|C_\ell(x) . h^\ell\|_p < \infty \tag{5.18}$$

and such that the cylindrical polynomials are dense in L^p. Then M is quasi-invariant with respect to all translations in H and

$$\frac{d(\tau_h M)}{d M} = \sum_{\ell=o}^{\infty} \frac{C_\ell(x) h^\ell}{\ell!} \in \cap L^p_M(X) \tag{5.19}$$

Proof. This sum converges in L^p by (5.18). Moreover for the arbitrary cylindriacl polynomial function F

$$< M , \tau_h F > = \int M (dx) F (x-h)$$

$$= \sum_{\ell=o}^{\infty}(-1)^\ell < \partial_h^\ell F , M > \ell!^{-1} = \sum_{\ell=o}^{\infty} < F , \partial_h^\ell M >$$

$$= \int F(x) \left[\sum_{\ell=o}^{\infty} \ell!^{-1} C_\ell(x) . h^\ell\right] M(dx) = < \tau_{-h} M , F >$$

(5.20) Weak divergence of cylindrical distribution. Putting $\mathcal{C}(X_\alpha , k) = \mathcal{C}(X_\alpha , \odot_k X'_\alpha)$ we have for

$X'_\beta \subseteq X'_\alpha \subseteq X'$ a canonical injection

$\mathcal{C}(X_\beta \, , \, k) \ni F \to (\bigodot_k i_{\alpha\beta}) \, F \, (s_{\alpha\beta}) \in \mathcal{C}(X_\alpha, k)$

Hence a space of vectorial and cylindrical test functions is defined on X putting

$$\mathcal{C}_{cyl}(X \, , \, k) = \lim_{\to} \mathcal{C}(X_\alpha \, , \, k) \qquad\qquad (5.21)$$

The weak divergence

$$L(\mathcal{C}_{cyl}(X \, , \, k+1) \, , \, V) \ni \to \delta \, T \tilde\in L \, (\mathcal{C}_{cyl}(X \, , \, k) \, , \, V) \qquad (5.22)$$

has been defined in [17-19] by

$$< \delta \, T \, , \, F > \; = \; < T \, , \, \nabla \, F > \qquad\qquad (5.23)$$

(5.24) Proposition. For any $G \in \mathcal{C}_{cyl}(X)$ and for any cylindrical measure $M = (M_\alpha)$ on X satisfying (5.4) and cylindrically smooth

$$\nabla^k (G \, M) \, . \, y^k = (\, B_k(G) \, . \, y^k) \, M \qquad\qquad (5.25)$$

with $B_k(G) \, . \, y^k = \sum_{\ell=o}^k \binom{k}{\ell} (\nabla^{k-\ell} G . y^{k-\ell})(C_\ell(x).y^\ell)$

Hence putting $\partial_y^k = (\partial_y)^k$ the following duality formula with $F \in \mathcal{C}_{cyl}(X)'$, $y \in X$

$$\int G(\partial_y^k F) \, M = < F \, , \, (B_k(G) \, . \, y^k) M > \qquad (5.26)$$

In the same way for arbitrary $Q \in \mathcal{C}_{cyl}(X, k')$ and $G \in \mathcal{C}_{cyl}(X)$

$$\delta^k (Q \, G) = \sum_{\ell=o}^k \binom{k}{\ell}(\delta^\ell \, Q) \, . \, (\nabla^{k-\ell} \, G) \qquad (5.27)$$

The proof uses the following formula   valid for arbitrary F and $G \in \mathcal{C}_{cyl}(X)$

$$G(\partial_y^k F) = \sum_{\ell=o}^k (-1)^{k-\ell} \binom{k}{\ell} \partial_y^\ell (F \, \partial_y^{k-\ell} \, G) \qquad (5.28)$$

This formula is proved by induction on k.

Proof of (5.25) and (5.26) :

$$((-1)^k < F , \partial_y^k (G \ M) > = - \int G(\partial_y^k \ F) \ M$$

$$= \sum_{\ell=0}^{k} (-1)^{k-\ell} \binom{k}{\ell} \int \partial_y^\ell (F(\partial_y^{k-\ell} \ G)) \ M$$

$$= (-1)^k \sum_{\ell=0}^{k} \binom{k}{\ell} F(\partial_y^{k-\ell} \ G)(\partial_t^\ell \ M) = (-1)^k < F , (B_k(G) . y^k)M >$$

and   (5.27)   is proved in the same way.

4. THE WEAK RELATIVE DIFFERENTIAL CALCULUS FOR MEASURES.

Application to the stochastic calculus :

Let H be a Hilbert subspace of a $\ell.c.H.s.$ X
The weak differential calculus relative to a given $H-C^\infty$
measure M on X is briefly presented below.

The image of the injective mapping

$$L_M^P(X) \ni F \longrightarrow (G \rightarrow \int F \ G \ M) \in \mathcal{C}_{cyl}(X)' \tag{6.1}$$

is denoted $L_M^P(X) \ M$. In the same way $\mathcal{C}_{cyl}(X) \ M$ can be
defined; hence a triplet

$$\mathcal{C}_{cyl}(X) \ M \xrightarrow{i} L^2(X) \ M \simeq (L^2(X) \ M)' \xrightarrow{i^T} \mathcal{C}_{cyl}(X)' \tag{6.2}$$

(6.3) Definition of the relative weak derivative $(\nabla^k F) \ M$
for any $F \in \cup L_M^P(X)$.

Since M is $H-C^\infty$, the following mapping is continuous
for any fixed h .

$$\mathcal{C}_{cyl}(X) \ni G \xrightarrow{B_k(.).h^k} B_k(G).h^k \in \cap L^P(X) \ M \tag{6.4}$$

Hence the duality formula (5.25) shows that the relative
derivation

$$\mathcal{C}_{cyl}(X) \ M \ni F \longrightarrow (\partial_h^k F) \ M \in \mathcal{C}_{cyl}(X) \ M \tag{6.5}$$

is extended by the transpose of (6.4) :

$$\underset{p>1}{\cup} L_M^P(X) \ M \xrightarrow{\partial_y^k} \mathcal{C}_{cyl}(X)' \tag{6.6}$$

More generally, for any $T \in L(L_M^P(X) \ M , V)$,

the relative derivative $\partial_h^k T \in L(\mathfrak{T}_{cyl}(X), V)$,

can be defined by

$$(\partial_h^k T)(F) = T(B_k(F).h^k) \tag{6.7}$$

The polynomial function $h \to \partial_h^k T$ is the global relative weak derivative of the cylindrical distribution T

(6.8) Using these preliminaries, Banach-valued Sobolev spaces with resp to the H-smooth measure M are defined as Hilbert valued Sobolev spaces have been defined in [17] [19] for the Gaussian case. As in [17] the study of the density of the subspace $\mathfrak{T}_{cyl}(X$ $W^{p.k}$ is reduced to the finite dimensional case using (5.5), and many different kinds of Sobolev spaces can be defined.

(6.9) Note that the Gaussian case is <u>very particular</u> since for any $X'_\alpha \Subset X$, M is the product of two measures. Therefore $B_k(.)y^k$ defines a linear and continuous mapping in $\mathfrak{T}_{cyl}(X)$ <u>for arbitrary</u> $y \in X$. Hence for any cylindrical distribution $T \in L(\mathfrak{T}_{cyl}(X), V)$, the relative weak derivative $\partial_y^k T$ can be defined in the Gaussian case.

(6.10) <u>Theorem 2</u>. (see [23] if g and g' are Gaussian)
  Let X and Y be two $\ell.c.H.s.$ Let H(resp K) be a dense Hilbert subspace of X(resp Y)
  Let g and g' be a H-smooth probability on X and a K-smooth probability on Y s.t. cylindrical test functions are dense in the Sobolev spaces with respect to g and g' resp. a) Then $gg' = g \otimes g'$ is H $\oplus$ K-smooth. b) Moreover, for arbitrary $p>2$, k and $\ell>0$ and for any Hilbert space G the following mapping is continuous and has norm $\leqslant (k+1)^{1/p}$

$$W^{p,k+\ell}(X \times Y, G) \ni F \longrightarrow \vec{F} = F(x,.) \in W^{p,k}(X, W^{2,\ell}(Y,G)) \tag{6.11}$$

The proof of a) uses an argument of topological tensor product. The proof of b) uses simply the Jensen inequality since putting $H'^i = \bigodot_i H'$ for $i \leqslant k$:

$$(\| \nabla^i \vec{F} \|_p)^p = \int_X g(dx) \left(\sum_{j=o}^{\ell} \int_Y \left| \nabla_x^i \nabla_y^j F(x,y) \right|^2_{H'^i \otimes K'^j \otimes G} g'(dy) \right)^{p/2}$$

$$\leqslant \sum_{j=o}^{\ell} \int g(dx) \int \left| \nabla_x^i \nabla_y^j F(x,y) \right|^p_{H'^i \otimes K'^j \otimes G} g'(dy)$$

$$\leqslant \sum_{j=0}^{\ell} \int g(dx) \int \left| \nabla^{i+j} F(x,y) \right|_{(H+K)^{\cdot}(i+j)\otimes G}^{p} g((dy))$$

$$\leqslant (\parallel F \parallel_{p,k+\ell})^{p}$$

(6.12) Now a $H - C^{\infty}$ probability measure P on X is considered s.t.

a) $\mathcal{C}_{cyl}(X)$ is dense in $W^{p,k}(X)$ for arbitrary $k > 0$ and $1 < p < \infty$

b) The divergence δ is continuous in the Hilbertian Sobolev spaces relative to P (see [15] in the Gaussian case; this is also a corollary of [24]

(6.13) Therefore a) by the Hölder inequality the Frechet space

$$W = \cap_{p,k} W^{p,k} (X)$$

is an algebra, b) in view of (6.12), (5.27) can be extended by continuity for $G \in W$ and $Q = F P$ with F in some vectorial Sobolev space : see [5] in the Gaussian case.

(6.14) Theorem 3. An Euclidean space Y is identified with the dual space and let ρ be the Riesz isometry of H' on the dual space $H'' = H$. Let $F \in W \otimes Y$ be such that ∇F is a.s. surjective $H \to Y$ and such that :

$$\gamma = \sigma^{-1} \in \cap_{p} L^{p}(X, \text{End } Y) \quad \text{with} \quad \sigma F = \nabla F \circ \rho \circ (\nabla F)^{T}$$

Then denoting m the law of F and putting $G^{F} = E [G \parallel F]$ the following linear mapping is continuous

$$W \ni G \longrightarrow G^{F} m \in \mathcal{G}(Y)dy \qquad (6.15)$$

(6.16) Corollaries : a) This gives in particular $m \in \mathcal{G}(Y)dy$ for $G = 1$: compare with [22] and see [16] in the Gaussian case. b) The transpose of (6.15) is the continuous mapping

$$S'(Y) \ni T \longrightarrow T \circ F \in W' \qquad (6.16)$$

(see [28] in the Gaussian case)

Proof. Let $(\varepsilon_i) = (\partial_i) = (\partial/\partial y_i)$ be an orthonormal basis of Y. Let $\mathrm{Diff}_p(Y)$ be the algebra of linear differential operators with polynomial coefficients on Y. A lifting of $d \subset \mathrm{Diff}_p(Y)$ by F is defined as any $D \in \mathrm{End}\ W$ such that :

$$E[\ (d\varphi)\ o\ F\ .\ G\ ]\ =\ E[\ (\varphi\ o\ F)\ D\ G\ ]$$

for arbitrary $\varphi \in \mathcal{C} = \overset{\alpha}{\mathcal{C}}(Y)$ and $G \in W$.

Denoting q the weight $1 + |y|^2$ on Y the following fact is well known for any probability measure m' on Y

$$\forall \ell \ \forall \alpha = (\alpha_1, \ldots \alpha_n) \quad q^\ell \partial^\alpha m' \text{ is a bd measure} \Rightarrow m' \in \mathcal{Y} \ dy$$

This can be applied to $m' = G^F m$. Since the conditionning induces a contraction in all L^P, the problem is to lift on X the differential operators : $\varphi \to d\varphi = \partial^\alpha (q^\ell \varphi)$ on Y. But the operator of product by q^ℓ on Y is lifted in the operator of product by $q^\ell(F) = q^\ell o F$ on X. And ∂_i is lifted in

$$G \longrightarrow D_i\ G = \delta((\gamma\ \varepsilon_i\ o\ \nabla F)\ G)$$

since [22]

$$E[\ (\partial_i \varphi)\ o\ F\ .\ G\ ] = E[\ <\ \nabla(\varphi o F), ((\gamma\ \varepsilon_i)\ o\ \nabla F)\ G\ ]$$

Therefore $d = (\partial_1)^{\alpha_1} (\partial_n)^{\alpha_n}(q^\ell_.)$ is lifted in

$$D = (q^\ell(F).)\ .\ (D_n^{\alpha_n})\ldots(D_1^{\alpha_1})$$

and the theorem 3 is proved.

Paul Krée Mathématiques

et LA n° 213 - Université Paris VI.

Place Jussieu - 75005 PARIS

REFERENCES.

[1]   S. Albeverio and R. Høegh-Krohn, Z. Wahrsch-Verw.
      Gebiete 40 (1977) pp.1-57.

[2]   P. Bernard, M. Fogli et C. Wagner. Actes des journées
      (Juin 1984) de mécanique aléatoire appliquées à la
      construction. Edited by AFREM and the Lab. Cent. des
      Ponts et Ch. (Paris) pp.157-165.

[3]   P. Bernard, P. Fogli, M. Bressolette, P. Lemaire
      To appear in Journal de Méc. Théor. et Appl.

[4]   R.F. Curtain and A.J. Pritchard Lect. Notes in Control
      and Inf. Science vol.8 (1978) Springer Verlag.

[5]   J. Diebolt. C.R. Acad. Sc. Paris t.296 (Juin 1983)
      Serie I. PP.837-840.

[6]   W.H. Fleming and L.G. Gorostiza Editors : Lecture
      Notes in Control and Information Sciences n°42.
      Springer Verlag 1982.

[7]   B. Gaveau. GR. Acad. Sc. Paris t.293 (Novembre 1981)
      Serie I. pp.469-472.

[8]   I.M. Gelfand and N.Y. Vilenkin. Les distributions,
      tome 4. Dunod Paris (1967).

[9]   J.L. Guilkman et A. Skorokhod. Introduction à la théorie
      des processus aléatoires. Ed. MIR Moscou. 1977-1980.

[10] R.Z. Hasminskii. Stochastic stability of differential
      equations. Sijthoff and Noordhoff (1980).

[11] M. Hazewinkel and J.C. Willems. Editors : Stochastic
      Systems ... Reidel (1981).

[12] K. Ito in Stochastic Analysis. North Holland 1984,
      pp.197-224.

[13] P. Krée and C. Soize. Mécanique aléatoire. Dunod Paris
      (1983) English translation in course by D. Reidel.

[14] P. Krée. Journ. of Math. Phys. 24(11) Novembre 1983.
      pp.2573-2580.

[15] M. and P. Krée. C.R. Acad. Sc. Paris t296 (Juin 1983)
      Serie I. pp.833-836.

[16] P. Krée. C.R. Acad. Sc. Paris. t.296 (Janvier 1983)
      Serie I. pp.223-225.

[17] P. Krée in Sem. P. Lelong I Lecture Notes in Mathema-
      tics (Springer Verlag) n°410 (1973).
      and II Lecture Notes in Mathematics n°474 (1974) pp.2.47

[18] P. Krée Journ. of Funct. Anal. Vol.31 n°2 (1979) pp.150-186

[19] P. Krée.Séminaire sur les ed.p en dim. infinie 1974-1975. Edited by Secrétariat math. of the H. Poincaré Institute (Paris)

[20] S. Kusuoka. J. Fac. Sci. Univ. Tokyo 29 (1982) pp.79-85.

[21] J.T. Lewis and L.C. Thomas. Wahrsch. Verw. Geb. 30, pp.45-55 (1974).

[22] P. Malliavin in Proc. Int. Symp. on SDE (1976) Kyoto ed. bt K. Ito Konokoniya, Tokyo (1978)

[23] P. Malliavin in Stochastic Analysis ed. by K. Ito North Holland (1984) pp.369-386.

[24] P.A. Meyer. Manuscript received by P. Malliavin in may 1983 and published in Sem. of Probability XVIII Lecture Notes n° 1059 pp. 179-193. Springer Verlag (1984).

[25] P.A. Meyer. The "Malliavin Calculus" and some pedagogy. to appear in Sem of Probability (Lect. Notes).

[26] P. Paclet. Expose 5 in Sém. P. Krée. 1977-1978. Edited by Secrétariat Math. of the H. Poincaré Institute (Paris).

[27] C. Soize. Actes des journées (Juin 1984, Paris) de mécanique aléatoire appliquée à la Construction. Edited by AFREM and the Lab. Cent. des Ponts et Ch. (Paris) pp.24-31.

[28] S. Watanabe. In Lect. Notes in Control and Inf. Science n°49 (1983) p.284-290, Springer Verlag. Berlin

A. S. USTUNEL

STOCHASTIC ANALYSIS ON NUCLEAR SPACES
AND ITS APPLICATIONS

Foreword

In this paper we review the recent developments in
the theory of stochastic processes with values in
the nuclear spaces. The choice of the subjects
follows our personal research and all the results
are announced without any proof. We have avoided
to give the detailed applications and tried to
give an overview of the subject in such a way that
the reader can use these results as soon as he
becomes familiar with the concepts illustrated
here.

Having explained why one needs a stochastic
calculus in the frame of the distributions, we
have given the basic definitions and some important
identification results in the first section. The
second section is devoted to the construction of
the stochastic calculus and some of its applications. In the third section we study in more detail
the trajectories of the weakly regular processes
and announce some hypoellipticity results for the
stochastic partial differential operators operating
on the distributions-valued semimartingales. The
weak convergence of the processes and the
characterization of the additive processes are also
given in this section.

As a conclusion all the results confirm the
following (probably hasty) conjecture: any
proposition which is true in finite dimension and
announcable on a nuclear space is true also there.

L. Arnold and P. Kotelenez (eds.), Stochastic Space-Time Models and Limit Theorems, 163–177.
© *1985 by D. Reidel Publishing Company.*

Motivations

Let W be an $\mathbb{R}^d$-valued   standard Wiener process, $T \in \mathcal{D}'(\mathbb{R}^d)$.
Define the function f on $\mathbb{R}^d$ by

$$f(\lambda) = \langle T, \varphi(\cdot + \lambda) \rangle \;,\; \varphi \in \mathcal{D}(\mathbb{R}^d).$$

Since f is a C^∞-function, Itô's formula gives

$$f(W_t) = T(\varphi) + \int_0^t \nabla f(W_s) \cdot dW_s + \tfrac{1}{2} \int_0^t \Delta f(W_s) ds$$

$$= T(\varphi) + \int_0^t \langle T, \partial_i \varphi(\cdot + W_s) \rangle \, dW_s^i + \tfrac{1}{2} \int_0^t \langle T, \Delta \varphi(\cdot + W_s) \rangle \, ds$$

Let us denote by $\tilde{X}(\varphi)_t$ the process $f(W_t)$. Then $\tilde{X}$ defines
a linear mapping from $\mathcal{D}(\mathbb{R}^d)$ into the space of the semimartingales (with respect to some filtration) such that

$$d\tilde{X}(\varphi)_t = \tfrac{1}{2} \tilde{X}(\Delta \varphi)_t \, dt + \tilde{X}(\partial_i \varphi)_t \, dW_t^i$$

$$\tilde{X}(\varphi)_0 = T(\varphi)$$

for any $\varphi \in \mathcal{D}(\mathbb{R}^d)$. A moment's reflexion suggests that there
exists some X : $\mathbb{R}_+ \times \Omega \mapsto \mathcal{D}'(\mathbb{R}^d)$ such that X has continuous
trajectories and $\langle X_\cdot, \varphi \rangle = \tilde{X}(\varphi)_\cdot$ for any $\varphi \in \mathcal{D}(\mathbb{R}^d)$. Hence
X is a "solution" of

(1) "$dX_t = 1/2 \; \Delta X_t \; dt - \; \partial_i X_t \; dW_t^i$ "

$$X_o = T$$

where the the integrals are to be defined. In fact what we
have done can be explained easily using a physical language:
we have two reference frames R1 and R2 such that R2 is
moving with respect to R1 following the Brownian path $W_\cdot(\omega)$:

and $T \in \mathcal{D}'_2(\mathbb{R}^d)$ is observed as $X_t(\varphi)$ by an observer in R1
at time t. Since $x \mapsto x + W_t(\omega)$ is reversible, the equation
(1) should have a "unique solution" and to prove this, it is
sufficient to show that

$$" X_t * \delta_{-W_t} = T "$$

where $\delta.$ is the Dirac measure whose support is "." and
"$*$" denotes the convolution .Of course,in order that this
discussion to be of some interest we should define first
what we understand by the words "solution","unique",etc.
In fact,all we did above can be rigourously explained using
the weak formalism except the calculation of the convolution:
for this,it is obvious that we need a kind of integration
by parts formula.This will be done by defining a class of
processes with values in $\mathcal{D}'(\mathbb{R}^d)$ with which it is possible
to construct a stochastic calculus.For the sake of general-
ity ,instead of working on $\mathcal{D}$ and $\mathcal{D}'$,we will work on
the more general spaces ,called nuclear spaces .

I.Notations and preliminaries

Φ denotes a complete,nuclear space whose continuous dual
Φ' is also nuclear under the strong topology $\beta(\Phi',\Phi)$,
denoted by Φ'_β . If U is an absolutely convex neighborhood
(of zero) in Φ ,we denote by $\Phi(U)$ the completion of
$\Phi / p_U^{-1}(0)$ with respect to the norm p_U,where p_U denotes
the gauge function of U; k(U) denotes the canonical mapping
from Φ into $\Phi(U)$. If $V \subset U$ is another such neighborhood
we define k(U,V): $\Phi(V) \longrightarrow \Phi(U)$ as $k(U)=k(U,V) \circ k(V)$.
Let us recall that in each nuclear space Φ , there exists
a neighborhood base $\mathcal{U}_h(\Phi)$ such that (cf. [10]) for any
$U \in \mathcal{U}_h(\Phi)$, $\Phi(U)$ is a separable Hilbert space and there
exists $V \in \mathcal{U}_h(\Phi)$, $V \subset U$ such that k(U,V) is a Hilbert-
Schmidt (or 2-nuclear) mapping.If B is an absolutely convex,
bounded,closed set, $\Phi[B]$ denotes the subspace spanned by B
equipped with the norm p_B (i.e.,the gauge function of B).
In this way , the continuous dual of $\Phi(U)$ can be identified
as $\Phi'[U°]$ where U° is the polar of U (cf.[10]) .

In the sequel we denote by $(\Omega, \mathcal{F}, \mathcal{F}_{t \geqslant 0}, P)$ a general probability space where $(\mathcal{F}_t, t \geqslant 0)$ is a right continuous, increasing filtration and $\mathcal{F}_0$ contains all the P-null sets.

Definition 1

i) Let X be the set $\{X^U; \ U \in \mathcal{U}_h(\bar{\Phi}'_\beta)\}$

where, for any U, X^U is a stochastic process with values in $\bar{\Phi}'(U)$. X is called a <u>projective system</u> (of stochastic processes) if, for any $V \in \mathcal{U}_h(\bar{\Phi}'_\beta)$, $V \subset U$, $k(U,V) \circ X^V$ and X^U are undistinguishable.

ii) X is called a (projective or generalized or g-) semi-martingale, martingale, etc., if for any $U \in \mathcal{U}_h(\bar{\Phi}'_\beta)$, X^U is respectively a semimartingale, martingale, etc., with values in $\bar{\Phi}'(U)$. More generally we say that X possesses the property π if X^U possesses the property π for any U in $\mathcal{U}_h(\bar{\Phi}'_\beta)$.

For $p \geqslant 1$, S^p denotes the space of real valued special semimartingales ($x = m + a$, $a_0 = 0$, a is previsible, of finite variation, m is a local martingale) equipped with the following norm

$$|x|_p = \left\| [m,m]_1^{1/2} + \int_0^1 |da_s| \right\|_{L^p(\Omega, \mathcal{F}, P)}$$

(we indexe our processes either with $[0,1]$ or $\mathbb{R}_+$ for the sake of notational simplicity). S^0 denotes the space of all real valued semimartingales equipped with its usual metrizable but non locally convex Fréchet topology.

In the following we shall prove to the reader that the definition 1 is in fact much more practical than it seems at first glance :

Theorem 1 (Identification Theorem, cf. [16], [17])

a) There exists a one to one correspondence between the pro-

jective semimartingales and the linear,sequentially contin-
ous mappings from $\bar{\Phi}$ into S° and any one of them is denoted
by $S^{\circ}(\bar{\Phi}')$.

b)There exists a one to one correspondence between S^p-pro-
jective semimartingales,the linear continuous mappings from
$\bar{\Phi}$ into S^p,the linear,sequentially continuous mappings from
$\bar{\Phi}$ into S^p and $\bar{\Phi}'_\beta \tilde{\otimes} S^p$,i.e., the completed projective
tensor product of S^p and $\bar{\Phi}'_\beta$, for $p \geqslant 1$ .This set will be
denoted by $S^p(\bar{\Phi}')$.

c)For any $X \in S^p(\bar{\Phi}'), p \geqslant 1$, there exists an absolutely convex
compact set $K \subset \bar{\Phi}'$,such that, $\bar{\Phi}'[K]$ is a separable Hilbert
space and a $\bar{\Phi}'[K]$ -valued S^p-semimartingale Y such that
$$\langle i_K(Y), \varphi \rangle = \tilde{X}(\varphi)$$
 for any $\varphi \in \bar{\Phi}$,where i_K denotes the canonical injection
$\bar{\Phi}'[K] \hookrightarrow \bar{\Phi}'$.

Remarks

1)The parts (b) and (c) are the consequences of the nuclearity,the closed graph theorem and the fact that $\bar{\Phi}$ is bornolo-
gical.

2) The part (a) is more difficult to prove because S^o is
not locally convex .

3)In case (c) we say that X lives in $\bar{\Phi}'[K]$.

   A special case is the following :

Theorem 2 (c.f. [19])

Suppose that $\bar{\Phi}'_\beta$ is metrizable .Then,for any $X \in S^{\circ}(\bar{\Phi}')$,
there exists a probability Q equivalent to P such that,
under Q , X is in $S^1(\bar{\Phi}')$ hence lives in some $\bar{\Phi}'[K]$
as a (P or Q) semimartingale.

Remark

The theorem is valid even for the projective systems having

right and left limits. Let us also note that this result
extends a recent result which says that in a nuclear, Fréchet
space any probability measure has a Hilbert support.

II. Construction of the stochastic calculus

Let H be a bounded, previsible process with values in Φ .
Then, there exists some $U \in \mathcal{U}_h(\dot{\Phi})$ such that H is bounded
and previsible in $\Phi[U^\circ]$. If $X \in S^\circ(\Phi')$, define

$$I(H)_t = \int_0^t < H_s, dX_s > = \int_0^t (H_s, dX_s^U)$$

the last integral is a Hilbert space stochastic integral.
Then $I(H)$ is well defined and it is independent of the choice
of U and this integral extends to the locally bounded
processes (cf. [18]) .

Integration by parts formula

If $X \in S^\circ(\Phi')$, $Z \in S^1(\Phi)$ then Z lives in some $\Phi[U^\circ]$,
$U \in \mathcal{U}_h(\dot{\Phi}_\beta)$. Hence, from the Hilbert space integration (cf.
[3]) we have

$$X(Z)_t = (X_t^U , Z_t)$$

$$= \int_0^t (X_{s-}^U, dZ_s) + \int_0^t (dX_s^U, Z_{s-}) + [\![X^U, Z]\!]_t$$

It is easy to show that the integrals are independent of
the choice of U (cf. [18]), hence $[\![X^U, Z]\!]$ is also independent
of the choice of U and is denoted by $[\![X, Z]\!]$. This result is
called integration by parts formula and we write it as

$$\tilde{X}(Z_t)_t = \langle X_t, Z_t \rangle = \int_0^t \langle X_{s-}, dZ_s \rangle + \int_0^t \langle dX_s, Z_{s-} \rangle + [\![X, Z]\!]_t$$

Some applications of stochastic calculus

Let z be an $\mathbb{R}^d$-valued semimartingale with $z_0 = 0$. Then, for
any $T \in \mathcal{D}'(\mathbb{R}^d)$, the following equation has a unique solution:

$$X_t = T - \int_0^t \partial_i X_{s-} \, dz_s^i + \frac{1}{2} \int_0^t \partial_{ij} X_{s-} \, d\langle z^{i,c}, z^{j,c} \rangle_s +$$

$$+ \sum_{0 < s \leq t} [X_s - X_{s-} + \sum_i \partial_i X_{s-} \cdot \Delta z_s^i]$$

In fact the solution is $T_* \delta_{z_t}$ and its uniqueness is straightforward with the use of the integration by parts formula.

For further applications we need the following regularity theorem which is a consequence of the Radon-Nikodym property of nuclear spaces :

Theorem 3 (Regularity Theorem, cf. [20])

Let X be a random field on $[0,1] \times \mathbb{R}^d$ such that $x \mapsto X(x)$ is C^∞ as an S^1-valued mapping. Then there exists an $\mathcal{E}(\mathbb{R}^d)$ -valued (i.e., the nuclear, Fréchet space of the infinitely differentiable functions on $\mathbb{R}^d$) semimartingale $\hat{X}$ such that, for any $x \in \mathbb{R}^d$, $X(x)$ and $\hat{X}(x)$ are undistinguishable.

As an application of this theorem we have

Corollary

Let $0 \subset \mathbb{R}^d$ be open and $b: \mathbb{R}^d \times 0 \to \mathbb{R}^d$, $G : \mathbb{R}^d \times 0 \mapsto \mathbb{R}^d \circ \mathbb{R}^d$ be C^∞-vector fields with bounded derivatives. Then, there exists an $\mathcal{E}(0) \tilde{\otimes} \mathbb{R}^d$ -valued semimartingale π such that, for any $z \in 0$, $\pi(z)$ is the solution of

$$dx_t(z) = b(x_t(z), z) \, dt + G(x_t(z), z) \, dW_t$$

$$x_0(z) = \pi_0(z) , \quad \pi_0 \in \mathcal{E}(0) \tilde{\otimes} \mathbb{R}^d,$$

where W is a d-dimensional Wiener process.

If z is a semimartingale with values in 0 (i.e., if δ_z is a semimartingale in $\mathcal{E}'(0)$: the distributions of compact support on 0), then the integration by parts formula tells us that

$$\{ \pi_t(\omega, z_t(\omega)) = \langle \delta_{z_t(\omega)} , \pi_t(\omega, \cdot) \rangle ; (t, \omega) \in \mathbb{R}_+ \times \Omega \}$$

is a semimartingale and it can be developped as

$$\pi_t(z_t) = \pi_o(z_o) + \int_o^t b(\pi_s(z_s), z_s) \, ds + \int_o^t \sigma(\pi_s(z_s), z_s) \, dW_s +$$

$$+ \int_o^t (D_z \pi_s)(z_{s-}) \, dz_s + \frac{1}{2} \int_o^t (D_z^2 \pi_s)(z_{s-}) \, d\langle z^c, z^c \rangle_s +$$

$$+ \int_o^t [D_z \sigma(\pi_s(z_s), z_s) + D_x \sigma(\pi_s(z_s), z_s) \, D_z \pi_s(z_s)] \, d\langle W, z^c \rangle_s +$$

$$+ \sum_{o < s \leq t}' [\pi_s(z_s) - \pi_s(z_{s-}) - D_z \pi_s(z_{s-}) \Delta z_s]$$

When the variable z is the initial condition of the diffusion and the semimartingale (z_t) is an Itô process then this formula is called Itô-Stratonovitch formula (cf.[12], [13], [14], [15], [19], [27]).In fact this result can be extended to the flows of the quasi-continuous semimartingales by the same method (cf.[2]) .

III.Some stochastic analysis

a)Continuity of the trajectories

In the definition of the projective processes we used the general nuclear spaces.In practice , all the nuclear spaces have some special properties which make the things easier. In this section we shall make a digression to study these cases.

Denote by R^o the space of the (equivalence classes) of right continuous,measurable,real valued processes having left limits with the following metric :

$$d(x,o) = E(\inf(1, \sup_t |x_t|)) \ .$$

Then we have the following

Theorem 4 (cf.[2])

Suppose that $J : \Phi \longrightarrow (R^o, d)$ is a continuous linear mapping and that Φ is separable.Then there exists a measurable process (X_t ; $t \in [0,1]$) with values in $\overline{\Phi}'$ having almost surely strongly right continuous trajectories with left

limits such that $\langle X_{\bullet}, \varphi \rangle$ belongs to the equivalence
class of $J(\varphi)$ for any $\varphi \in \Phi$. Furthermore,there exists an
increasing sequence of absolutely convex,compact subsets
(K_n) of Φ' such that almost surely the process X lives
in $\bigcup_{n \in \mathbb{N}} \Phi'[K_n]$.

Remarks

1) For $\Phi' = \mathcal{J}'$ this result has been proved in [5] (evi-
dently without the second assertion of the theorem since
$\mathcal{J}'$ itself is a countable union of a sequence of compact
increasing sets),the extension to general $\overline{\Phi}'$ and the
fact that the trajectories are concentrated in a sequence
of Hilbert spaces have been proved in [2] . Moreover [2]
gives an Itô's formula for the semimartingales in nuclear
spaces with the help of this result .

2) In practice all the nuclear that we use are separable,
therefore the hypothesis of separability is not a
restriction. To verify the continuity of J the easiest
method is to show first its sequential continuity ,then, if
Φ is a Fréchet space or a countable inductive limit of
Fréchet spaces (this is always the case in applications)
this implies the continuity of J (cf.[1]) .

3) We have fixed a metric on R° but this is not obligatory,
any distance compatible with its vector space structure
would be acceptable.

b) Applications to SPDE

In [26] we have applied the Theorem 4 to show the hypoellipticity of the stochastic partial differential operators :
We want to study the equation

(1) $du_t = -pu_t \, dt + q_i u_t \, dW_t^i + dh_t$

where $n \in S^{\bullet}(\mathcal{E}(0)), 0 \subset \mathbb{R}^d$ open, W is an $\mathbb{R}^n$-Wiener process,
$p=p(t,\omega,x,\partial_x), q_i=q_i(t,\omega,x,\partial_x)$, degree $p=2m$, degree $q_i=m_i \leqslant m$;
and the coefficients of p and q_i are supposed to be $\cdot C^{\infty}$
with uniformly bounded derivatives on the compact sets
(they are local operators !) .

Definition

We say that the equation (1) is hypoelliptic if any solution
u in $S^{\bullet}(\mathcal{D}'(0))$ can be modified as an $\mathcal{E}(0)$-valued semimartingale given $h \in S^{\bullet}(\mathcal{E}(0))$.

Let us denote by A_2 the space of the semimartingales k
such that

$$k_t = k_0 + \int_0^t D_+ k_s \, ds + \int_0^t \partial_{W^i} k_s \, dW^i_s$$

with

$$E \int_0^t \left[|D_+ k_s|^2 + \sum_i |\partial_{W^i} k_s|^2 \right] ds < +\infty \quad , \quad t \in \mathbb{R}_+.$$

Now we can announce :

Theorem 5 (Hypoellipticity Theorem; cf. [26])

Suppose that there exists some $s > 0$ such that for any
$K \subset\subset 0$,for any bounded stopping time T, there exists $c, \bar{c} > 0$
depending only on K, s and T such that

$$\| u \|^2_{m+s-1,T} = E \int_0^T \| u_r \|^2_{m+s-1} \, dr \leqslant$$

$$\leqslant c\left(B_T(u,u) + \bar{c} \| u \|^2_{0,T} \right)$$

for any $u \in A_2 \tilde{\cap} \mathcal{D}_K(0)$, where $\| \cdot \|_\alpha$ denotes the Sobolev
norm of order α , $\mathcal{D}_K(0)$ is the space of C^{∞}-functions
on 0 whose supports are in K (compact) , B_T is the bilinear
form defined by

$$B_T(u,v) = E \int_0^T ((D_+ + p_r)u_r, v_r)_0 \, dr + \frac{1}{2} E \sum_{i=1}^n \int_0^T (\partial_{W^i} u_r, \partial_{W^i} v_r) \, dr$$

and $(.,.)_o$ denotes the $L^2(0,dx)$-scalar product. Then the
following stochastic partial differential equation is
hypoelliptic :

$$du_t = (-p + ((-1)^{m_i+1} / 2) \ q_i^2) \ u_t \ dt + q_i u_t \ dW_t^i + dn_t .$$

Remarks

1) For the proof of this result we refer to $[26]$.

2) Note that these kind of equations have always been studied

in the frame of Sobolev spaces and our approach ,as far as

we know, is the first treatment of the equation in its

natural setting,i.e. , as an equation in $S^o(\mathcal{D}'(\mathbb{R}^d))$.

c) Weak convergence of the processes

If $(X^n ; n \in \mathbb{N})$ is a sequence of projective systems of
continuous processes,then,we can define the weak covergence
or convergence in law as the convergence of $(P_n^U ; n \in \mathbb{N})$
on $\mathcal{Q}^U = C([o,1], \Phi'(U)) , U \in \mathcal{U}_h(\Phi_\beta')$. We have in fact

Theorem 6 (cf.$[6]$)

The sequence $(P_n ; n \in \mathbb{N})$ of projective probabilities is
tight if and only if ,for any $\varphi \in \Phi$, $(\tilde{X}_n(\varphi) ; n \in \mathbb{N})$ is
tight.

   Suppose that Φ is a nuclear Fréchet space or inductive
limit of a sequence of such spaces . Then (it is separable)
as we have explained in (b),the projective systems of
continuous or right continuous processes coincide with the
strongly continuous or right continuous processes.Hence,
for any $\{ P^U ; U \in \mathcal{U}_h(\Phi_\beta') \}$ on $\{ C([o,1], \Phi'(U)); U \in \mathcal{U}_h(\Phi) \}$, there
exists a (Radon) probability measure P on $C([o,1], \Phi_\beta')$ such
that $k'(U)(P)=P^U$ where $k'(U)$ is defined in an obvious way
from $k(U)$. Let us announce a tightness result in this case:

Theorem 7

Suppose that $\overline{\Phi}$ is a countable inductive limit of Fréchet spaces.Then a sequence of probabilities $(P_n ; n \in \mathbb{N})$ on $C([o,1],\overline{\Phi}'_\beta)$ is tight if and only if $(x^n_\cdot(\varphi) ; n \in \mathbb{N})$ is tight in $C([o,1], \mathbb{R})$ for any $\varphi \in \overline{\Phi}$.

Remarks :

1) These results have not been found in the above order . In fact Theorem 7 was proved in [5] for the space of the tempered distributions, the Theorem 6 is proved in [6], moreover one can find there the case where $\overline{\Phi}$ is a strict inductive limit of a sequence of Fréchet spaces.Relaxing the strictness hypothesis is then straightforward .

2)Of course all these results are true for the right contin- uous processes with left limits.

d)Additive processes

The characterization of additive processes on nuclear spaces is as easy as in the finite dimensional case.In the following we shall announce the results for the ordinary sense processes (not for the projective systems).Hence we suppose that Φ is separable and Souslin :

Theorem 8 (cf. [22])

Suppose that we are given a functional f_t on $\overline{\Phi}$ for $t \in [o,T]$, $T>o$, of the following form :

$$f_t(\varphi) = \exp \left(ia_t(\varphi) - (1/2) \mathfrak{b}_t(\varphi,\varphi) + \int_{\Phi'} [\exp i x(\varphi) - 1 - i x(\varphi)] \mu_t(dx) \right)$$

such that

i)$t \mapsto a_t$ is continuous from $[o,T]$ into $\overline{\Phi}'_\beta$

ii) $t \mapsto \mathfrak{b}_t$ is continuous uniformly on the compact sets of $\overline{\Phi} \times \overline{\Phi}$,with values in the space of the bilinear forms on $\overline{\Phi}$, for any t, $\mathfrak{b}_t$ is nonnegative definite.

iii) $\mu_t, t \in [o,T]$, is an abstract set function on the measurable subsets of $\bar{\Phi}'$,increasing with t continuously and there exists an absolutely convex compact $K \subset \bar{\Phi}'$ such that $\mu_t(K^c \cap \bullet)$ is a Radon measure on $\bar{\Phi}'$ for any $t \geqslant o$ with

$$\int_K p(x)^2 \, \mu_t(dx) < + \infty$$

for any   continuous seminorm p on $\bar{\Phi}'$.

Then there exists an additive process X having right continuous trajectories with left limits in Φ'_β whose characteristic function is given by f_t. Conversely to any such additive process X, corresponds a functional f_t ,$t \in [o,T]$,described as above .

Remarks

1)From Theorem 1 , X is a semimartingale if and only if the deterministic process $t \mapsto a_t^{(\varphi)}$ is of finite variation for any $\varphi \in \bar{\Phi}$.

2)Such an additive process can be decomposed as the sum of a deterministic process,a Gaussian process and a jump process as in the finite dimensional case (ie.,Paul Lévy's Theorem holds)(cf. [22]) .

A. S. Ustunel
2, Bd. A. Blanqui, 75013
Paris, France

REFERENCES

[1] Adasch,N.,Ernst,B.and Keim,D(1978).Topological Vector
    Spaces.Lect.N. in Math.Vol.639.Springer.

[2] Martias,C.(1983).Sur les supports des processus à valeurs
    dans des espaces nucléaires.Preprint.

[3] Métivier,M.&Pellaumail,J.(1980).Stochastic Integration.
    Academic Press.

[4] Mitoma,I.(1983).On the sample conjunity of S'-processes.
    J.Math.Soc.Jap.Vol.35,no.4.

[5] Mitoma,I.(1984).Tightness of probabilities on C([o,1],S').
    and on D([o,1],S').Annals of Probability .

[6] Fouque,J.P.(1984).La convergence en loi pour les processus
    à valeurs dans un espace nucléaire.A.I.H.P.,Vol.20,p.225.

[7] Nelson,E.(1967).Dynamical Theories of Brownian Motion.
    Princeton University Press.

[8] Pardoux,E.(1982).Equations du filtrage nonlinéaire de la
    prédiction et du lissage.Stochastics,Vol.6,p.193-231.

[9] Chaleyat-Maurel,M.&Michel,D.(1984).Hypoellipticity theo-
    rems and conditional laws.Z.f.W.

[10]Schaefer,H.H.(1971).Topological Vector Spaces.Grad.Texts
    in Math.Springer.

[11]Schwartz,L.(1973).Théorie des Distributions.3rd.Ed.Hermann
                                                      Paris.

[12]Bismut,J-M.(1981).A generalized formula of Itô and other
    properties of stochastic flows.Z.f.W.,55,331-350.

[13] Bismut,J.M.(1981).Mécanique Aléatoire.Lect.N.in Math.
     Vol.866,Springer.

[14] Kunita,H.(1981).Some extensions of Itô's formula.Sém.à
     de Prob.XV,p.118-141.Springer.

[15] Rozovskii,B.L.(1973).A formula of Itô-Ventcell.Vestnik
     Moskou Univ. ,Ser.1 Math.Mch.26-32.

[16] Ustunel,A.S.(1982).A characterization of semimartingales
        on nuclear spaces.Z.F.W.,60,21-39.

[17] Ustunel,A.S.(1983).An erratum to "A characterization of
        semimartingales on nuclear spaces" after Laurent
        Schwartz.To appear in Z.f.W.

[18] Ustunel,A.S.(1982).Stochastic integration on nuclear
        spaces and its applications.A.I.H.P.,18,no.2,165-200.

[19] Ustunel,A.S.(1982).Some applications of stochastic in-
        tegration in infinite dimensions.Stochastics,7,255-288.

[20] Ustunel,A.S.(1983).Analytic semimartingales and their
        boundary values.J.F.A.,Vol.51,no.2,p.142-158.

[21] Ustunel,A.S.(1983).Distributions-valued semimartingales
        and applications to control and filtering.Lect.N.in
        Cont.and Inf.Sci.Vol.61,p.314-325.Springer.

[22] Ustunel,A.S.(1984).Additive processes on nuclear spaces.
        The Annals of Prob.Vol.12,no.3,p.858-868.

[23] Ustunel,A.S.(1982).A generalization of Itô's formula.
        J.F.A.,47,143-152.

[24] Ustunel,A.S.(1983).Stochastic Feynman-Kac formula.J.
        d'Analy.Math.Vol.42,155-165.

[25] Ustunel,A.S.(1983).Some applications of stochastic
        calculus on nuclear spaces to the nonlinear
        stochastic problems.Nato ASI,Reidel Publ.Co.

[26] Ustunel,A.S.(1984).Hypoellipticity of the stochastic
        partial differential operators.Preprint.

[27] Ventzell,A.D.(1965).On equations of the theory of con-
        ditional processes.Theor.Prob.and Appl.10,357-361.

C. VAN DEN BROECK

STOCHASTIC LIMIT THEOREMS : SOME EXAMPLES FROM
NONEQUILIBRIUM PHYSICS.

1. INTRODUCTION.

Fluctuations are the deviations of variables from their average value. Due to the discrete nature of the processes underlying the macroscopic phenomena, such fluctuations are present
in every physico-chemical system. In many cases, it is found
that these fluctuations are small. For instance, the dispersion
of the number of gasparticles $X(V)$ in a volume V is
typically (Landau and Lifschitz, 1967)

$$<\delta X^2 (V)> \approx <X(V)> \sim V \qquad (1)$$

We conclude that the amplitude of the fluctuations, $<\delta X^2 (V)>^{1/2}$,
is very small as compared to the average value $<X(V)>$ in a
macroscopic system, i.e. for V large. Analogous results are
found for the mean square displacement of a Brownian particle,
along the x-axis, in function of time (D is the molecular
diffusion coefficient) (Einstein, 1905)

$$<\delta x^2 (t)> = 2Dt \sim t \qquad (2)$$

and for the light intensity I, scattered by N randomly distributed particles (E is the electrical field) (Berne and Pecora,
1976)

$$I = <E^2> \sim N \qquad (3)$$

From these results, one should not conclude that fluctuations
are unimportant on the macroscopic scale. Firstly, some phenomena, whose effects can be observed macroscopically, are enti-
rely due to the fluctuations, e.g. Brownian motion, light
scattering, the escape of a thermally agitated particle out
of a potential well or longitudinal dispersion (see section
4). Secondly, it is not always true that fluctuations are
small. Large fluctuations can be associated to long range
coherent behavior in a system close to a phase transition

179

L. Arnold and P. Kotelenez (eds.), Stochastic Space-Time Models and Limit Theorems, 179–189.
© 1985 by D. Reidel Publishing Company.

point (see sections 2 and 3). This particular behaviour is
of interest to the physicist and mathematician alike. For
the physicist, it is valuable to understand the mechanisms
by which a system can undergo a transition from one macroscopic state to another one. For instance, it turns out that
this mechanism is fundamentally different for a nonequilibrium
transition towards a self-organized state and for a transition
between different equilibrium structures, although the mathematical difficulties encountered are of the same type. These
difficulties are due to the breakdown of the central limit
theorem (in the sense that a probability density can not be
approximated by a Gaussian probability density) : close to
the transition point, the macroscopic quantities are no longer
the result of a large number of "weakly correlated" microscopic contributions. Physicists have deviced a number of
more or less successful techniques to calculate relevant
features for these critical situations (see e.g. Ma, 1975).
In doing so, they rely on physical argumentation rather than
on rigorous mathematical results, which are rather scarce.
In section 2, we briefly review the Van Kampen expansion of
the master equation. In section 3, we discuss the breakdown
of this expansion and propose an alternative "critical expansion". In section 4, we discuss a central limit theorem for
the problem of longitudinal dispersion.

2. THE VAN KAMPEN EXPANSION OF THE MASTER EQUATION.

Let us consider a macroscopic system characterized by an extensive variable X, e.g. the total number of particles in a given
total volume V. Such a variable is the sum of a large number of "microscopic contributions" $X_{\underset{\sim}{r}}$:

$$X = \sum_{\underset{\sim}{r}} X_{\underset{\sim}{r}} \tag{4}$$

i.e. $X_{\underset{\sim}{r}}$ is the number of particles in a small subvolume ΔV
centered at the position vector $\underset{\sim}{r}$; the sum over $\underset{\sim}{r}$ runs over
all the subvolumes constituting the total volume V. Generally
speaking, the microscopic quantities $X_{\underset{\sim}{r}}$ will be "strongly
fluctuating", i.e. $<\delta X_{\underset{\sim}{r}}^2>^{1/2}$ is of the same order of magnitude
as $<X_{\underset{\sim}{r}}>$ (e.g. the number of particles in a very small box
will fluctuate between 0 and 1). On the other hand, we expect
for the case of short range interaction forces between the
particles that the spatial correlations are equally short
ranged : $<\delta X_{\underset{\sim}{r}} \delta X_{\underset{\sim}{r}'}> \simeq 0$ for $|\underset{\sim}{r} - \underset{\sim}{r}'| > l_c$, where l_c is the so-

called correlation length. Hence, if one chooses subvolumes ΔV with linear dimension larger then l_c, one has

$$<\delta X_{\underset{\sim}{r}} \delta X_{\underset{\sim}{r}'}> \simeq <\delta X^2_{\underset{\sim}{r}}> \delta^{Kr}_{\underset{\sim}{r},\underset{\sim}{r}'}$$

If the system has translational invariance, then $<\delta X^2_{\underset{\sim}{r}}>$ is a constant and one concludes (see also eq. (1)) :

$$<\delta X^2> = \sum_{\underset{\sim}{r}} \sum_{\underset{\sim}{r}'} <\delta X_{\underset{\sim}{r}} \delta X_{\underset{\sim}{r}'}> = <\delta X^2_{\underset{\sim}{r}}> \frac{V}{\Delta V} \sim V \qquad (5)$$

By the same token , one expects that the probability distribution, for X will be asymptotically, for large V, a Gaussian distribution.

Let us apply the above ideas to the case of an extensive time dependent random variable X(t). We expect that one can write :

$$X(t) = \bar{X}(t) + X(t) \qquad (6)$$

where $\bar{X}(t)$ is the macroscopic part, proportional to the extensivity parameter V, and $\delta X(t)$ are small deviations of order $V^{1/2}$. This idea was exploited by Van Kampen (1961) to set up an asymptotic expansion for the probability distribution P(X,t) of a Markov process describing a physico-chemical system. For sake of simplicity, we consider the case of a Markov process X(t) taking values in ℕ. The conditional probability $P(X,t/X_0,t_0)$ obeys the following birth and death master equation :

$$\partial_t P(X,t/X_0,t_0) = \lambda(X-1)P(X-1,t/X_0,t_0)$$

$$+ \mu(X+1)P(X+1,t/X_0,t_0) - [\lambda(X)+\mu(X)]P(X,t/X_0,t_0) \qquad (7)$$

The extensivity property is secured by the extensivity of the birth and death rates $\lambda \sim V$ and $\mu \sim V$. We are interested in the behavior of X(t) for large volume size $V \to \infty$. Van Kampen (1961) showed that, in this limit, the stochastic process X(t) converges for finite times, to a deterministic process $\bar{X}(t) = \bar{x}(t)$. V, obeying the following ordinary differential equation

$$\frac{d\bar{x}(t)}{dt} = f(\bar{x}(t)) = \lim_{V \to \infty} \frac{\lambda(\bar{X}(t))-\mu(\bar{X}(t))}{V} \qquad (8)$$

and that the scaled stochastic process :

$$u(t) = \frac{X(t) - \overline{X}(t)}{V^{1/2}} \tag{9}$$

converges, for finite times, to an Ornstein Uhlenbeck process obeying the following stochastic differential equation

$$du(t) = f'(\overline{x}(t)u(t) + Q^{1/2}dW(t) \tag{10}$$

f' stands for the derivative of f with respect to $\overline{x}$, W(t) is the Wiener process and $Q = Q(\overline{x}(t))$ is the noise intensity, given by

$$Q(\overline{x}(t)) = \lim_{V \to \infty} \frac{\lambda(\overline{X}(t)) + \mu(\overline{X}(t))}{V} \tag{11}$$

For a rigorous formulation and proof of these theorems, see Kurz (1972). As far as the approach of the stationary state, i.e. the limit t → +∞, is concerned, one has to specify the stability properties of the macroscopic trajectory $\overline{x}(t)$. For instance, in the case of a bistable system, the steady state equation

$$f(\overline{x}^{st}) = 0 \tag{12}$$

possesses three solutions, two of which are asymptotically stable. In this case, the fluctuations corresponding to the transitions between these two states are of macroscopic order of magnitude, $<\delta X^2> \sim V^2$, and the convergence of the stochastic trajectory to the macroscopic one breaks down. A problem of particular interest is the onset of bistability at the so-called cusp bifurcation point. In this case, the unique stationary state is asymptotically but not linearly stable, i.e.

$$f(\overline{x}^{st}) = f'(\overline{x}^{st}) = f''(\overline{x}^{st}) = 0 \; ; \; f'''(\overline{x}^{st}) < 0 \tag{13}$$

Since this situation lies at the borderline of the region of bistability ($<\delta X^2> \sim V^2$) and the "Gaussian regime" ($<\delta X^2> \sim V$, for all times, before bifurcation), we expect anomalous fluctuations, $<\delta X^2> \sim V^{2a}$ with 0.5 < a < 1. These fluctuations can be studied by considering the scaled stochastic process

$$u'(t) = \frac{X(t) - \overline{X}(t)}{V^a} \tag{14}$$

The exponent a is choosen such that the resulting process
u'(t) is nontrivial in the limit $V \to \infty$: if a is choosen too
small, e.g. a = 1/2 as in eq. (9), one will find

$$\lim_{t\to+\infty} <u'^2(t)> = +\infty$$

if a is taken too large, one obtains

$$\lim_{t\to\infty} <u'^2(t)> = 0$$

Instead of pursuing these ideas dealing with "zero dimensional systems" (one scalar variable X), we will turn to the
much more involved case of spatially distributed systems. Our
motivation is twofold. Firstly, as it is clear from the arguments given at the beginning of this section, the breakdown
of the result $<\delta X^2> \sim V$ must be associated to the arisal of
long range spatial correlations. This is not surprising since
we expect that the onset of new macroscopic states (which in
more complicated cases exhibit spatial and/or temporal orde-
ning) must involve long range coherence throughout the system.
Secondly, the homogeneous description considered above is a
mean field type of approximation since, in such a description the effect of inhomogeneous fluctuations is neglec-
ted. Such a description has been shown to lead to qualitati-
vely wrong results in the theory of equilibrium phase transition (Ma, 1975).

3. CRITICALITY EXPANSION OF THE MULTIVARIATE MASTER EQUATION.

In order to take into account the effect of inhomogeneous fluctuations, it is necessary to formulate a master equation approach for spatially distributed system. We partition the
system into an array of cubic subcells, centered at the position vectors $\{r\}$. The basic assumption is to suppose that the
set of the particle numbers in these cells, $\{X_r\}$, constitute
a Markov process. For the so-called reaction diffusion systems, two mechanisms are responsible for the time evolution
of the multivariate probability distribution $P(\{X_r\},t)$: the
chemical reactions, taking place inside each cell, being modelled by a birth and death process (cf. the previous section)
and diffusion between adjacent cells, described by a random
walk. As an example, we mention the following model (C. Van
den Broeck et alii, 1980) :

$$X \xrightarrow{k_1} B \qquad A \xrightarrow{k_2} 2X \tag{15}$$

k_1 and k_2 are chemical rate constants and A is further used to denote the fixed number of particles, per cell, of a controlled chemical species. In terms of the generating function :

$$F(\{s_{\underset{\sim}{r}}\},t) = \sum_{\underset{\sim}{r}} \sum_{X_{\underset{\sim}{r}}=0}^{\infty} \prod_{\underset{\sim}{r}'} s_{\underset{\sim}{r}'}^{X_{\underset{\sim}{r}'}} P(\{X_{\underset{\sim}{r}}\},t) \tag{16}$$

the multivariate master equation takes the following form :

$$\partial_t F(\{s_{\underset{\sim}{r}}\},t) = \sum_{\underset{\sim}{r}} [-k_1 (s_{\underset{\sim}{r}} - 1) \frac{\partial}{\partial s_{\underset{\sim}{r}}} \tag{17}$$

$$+ k_2 A(s_{\underset{\sim}{r}}-1)(s_{\underset{\sim}{r}}+1) + D \sum_{\underset{\sim}{1}} (s_{\underset{\sim}{r+1}} - s_{\underset{\sim}{r}}) \frac{\partial}{\partial s_{\underset{\sim}{r}}} F(\{s_{\underset{\sim}{r}}\},t)$$

The sum over $\underset{\sim}{1}$ runs over all the nearest neighbours and D, the jump rate, is related to Fick's diffusion coefficient D_F (which is a constant, independent of the cell size) by :

$$D = \frac{D_F}{(\Delta V)^{2/d}} \tag{18}$$

d is the dimensionality of the system.

Generally speaking, the solution of a multivariate master equation is very complicated, due to the large number of variables. Moreover, the cell size ΔV is not an obvious perturbation parameter, since it must be choosen small enough to describe the effect of inhomogeneous fluctuations. Rigorous results have been obtained by L. Arnold and P. Kotelenez (1983), but they are restricted to the case of linear reactions For the more complicated model (15), we mention the instructive exact result for the equal time steady state pair correlation function (d = 3, $x_{\underset{\sim}{r}}(t) = X_{\underset{\sim}{r}}(t)/\Delta V$)

$$\lim_{\Delta V \to 0} \lim_{t \to \infty} = \frac{<\delta X_{\underset{\sim}{r}}(t)\delta X_{\underset{\sim}{r}'}(t)>-<X_{\underset{\sim}{r}}(t)>\delta_{\underset{\sim}{r},\underset{\sim}{r}'}^{Kr}}{\Delta V^2}$$

$$= <\delta x_{\underset{\sim}{r}} \delta x_{\underset{\sim}{r}'}>^{st} - <x_{\underset{\sim}{r}}>^{st} \delta(\underset{\sim}{r}-\underset{\sim}{r}') = \frac{\phi}{2\pi |\underset{\sim}{r}-\underset{\sim}{r}'| D_F} \exp(-|\underset{\sim}{r}-\underset{\sim}{r}'|/l_c)$$

$$(19)$$

$\phi = k_2 A$ is the material flux through the system. The correlation length l_c is given by :

$$l_c = (D_F/k_1)^{1/2} \tag{20}$$

We conclude that the correlation length is of the order of the distance, travelled by a particle between two reactive collisions. This result suggests that, for a more general reaction scheme, the correlation length will be of the form (20) with k_1 replaced by the total linearized decay rate $|f(\bar{x}^{st})|$ (cf. eq. 10), which we will further denote by δ :

$$l_c = (D_F/\delta)^{1/2} \sim \delta^{-\nu} \tag{21}$$

Note the divergence of the correlation length with a so-called classical critical exponent $\nu = 1/2$, at the approach of the bifurcation point $\delta = 0$. This divergence is responsible for the anomalous behavior of the global fluctuations

$$\lim_{V \to \infty} \frac{<\delta X^2>}{V} \sim \delta^{-\gamma} \qquad \text{with} \quad \gamma = 1 \text{ and } X = \int_V x_{\underset{\sim}{r}} d\underset{\sim}{r}$$

To check the validity of the result (21), a systematic expansion of the master equation was set up. The basic features of this criticality expansion are as follows :

- We anticipate the existence of long range spatial correlations. Small scale fluctuations are thus expected to be negligible and one can blow up the cell size ΔV. The number of cells is kept constant.
- By keeping the dominant terms in $\varepsilon = 1/\Delta V$ in the multivariate master equation, one shows that the discrete Markov process $X_{\underset{\sim}{r}}$, converges to a continuous Markov process $x_{\underset{\sim}{r}}.\Delta V$, where $x_{\underset{\sim}{r}}$ obeys the following Langevin equation with nonlinear drift but process independent noise

$$\frac{\partial x_{\underset{\sim}{r}}}{\partial t} = f(x_{\underset{\sim}{r}}) + D \sum_{\underset{\sim}{1}} (x_{\underset{\sim}{r}+\underset{\sim}{1}} - x_{\underset{\sim}{r}}) + \varepsilon^{1/2} F_{\underset{\sim}{r}}(t) \tag{22}$$

where $F_{\underline{r}}$ is a multi-Gaussian white noise with correlation
function :

$$\langle F_{\underline{r}}(t)F_{\underline{r}'}(t')\rangle = \delta(t-t')\{Q(\overline{x}_{\underline{r}})\delta_{\underline{r},\underline{r}'}^{Kr}$$

$$+ D \sum_{\underline{1}} [(\overline{x}_{\underline{r}+\underline{1}}-\overline{x}_{\underline{r}})\delta_{\underline{r},\underline{r}'}^{Kr} - (\delta_{\underline{r}+\underline{1},\underline{r}'}^{Kr} - \delta_{\underline{r},\underline{r}'}^{Kr})(\overline{x}_{\underline{r}}+\overline{x}_{\underline{r}'})]\} \qquad (23)$$

The function f and Q are defined by the eqs. (8) and (11).
We have thus established a central limit theorem, ensuring
the convergence of a discrete Markov process to a continuous
one. The particular status of this theorem is revealed when
we calculate from it the stationary probability distribution.
For the simplest example ($f(x) = -\delta x - x^3$, Q = 1, this
corresponds to the Schögl model), one finds :

$$P_{st}(\{x_{\underline{r}}\}) = \exp\{\frac{2\epsilon}{Q_{st}}$$

$$\sum_{\underline{r}} [\frac{\delta}{2} (x_{\underline{r}}-\overline{x}_{st})^2 + \frac{1}{4} (x_{\underline{r}}-\overline{x}_{st})^4 + \frac{D}{4} \sum_{\underline{1}} (x_{\underline{r}+\underline{1}}-x_{\underline{r}})^2]\} \qquad (24)$$

This probability does not belong to the class of infinitely
divisible laws, even in the homogeneous case (D = 0), due to
the presence of a quartic term in the exponential. This re-
flects the coherence associated with bifurcation : the system
can no longer be partitioned into a collection of weakly correlated subsystems. On the other hand the exponential in (24)
displays the familar Landau Ginzburg type of potential in
discrete (position) space. This establishes the connection
between the Master Equation approach and the theory of equilibrium critical phenomena. Relevant properties, such as the
spatial correlation, critical dimensionality and critical
exponents can be obtained from renormalization group theory
(Ma, 1975). In particular, the correlation length is diver-
gent at the approach of the bifurcation point for dimensions
d > 1, validating our perturbative expansion in these cases.
For d = 1, the system exhibits no critical behaviour at the
crossing of the bifurcation point. The divergence of the correlation on length l_c in d > 1, is not always characterized by
the classical critical exponent $\nu = 1/2$. For instance, for
the case of a probability distribution of the type (24),
one has $\nu = 1$ and $\gamma = 1.75$ for d = 2 and $\nu \approx 0.64$ and $\gamma \approx 1.25$
in d = 3. Only for a dimension larger then the so-called

critical dimensionality d_c = 4, one recovers the classical
critical exponents ν = 1/2 and γ = 1 (Ma, 1975).

4. A CENTRAL LIMIT THEOREM FOR LONGITUDINAL DISPERSION.

In the previous sections, we discussed the stochastic properties of an extensive variable X in the limit of a large volume
size V. In this section, we will illustrate how analogous results are obtained for the case of layered systems in the long
time limit.
    We consider a system, consisting of n parallel horizontal
layers i = 1,..., n. In each of these layers, a fluid is flo-
wing with a velocity u_i, along the horizontal x-axis. We
follow the stochastic trajectory of a suspended particle,
which performs a random walk in the vertical direction over
the n layers, while being carried along with the local fluid
flow, in each layer. The transition probabilities, per unit
time, from layer i to i $\pm$ 1 are $k_i^{\pm}$. For simplicity of notation, we set $k_{-1}^+ = k_{n+1}^- = 0$, while $k_1^- = k_n^+ = 0$. The probability
densities P(x,i,t) to find the particle at the position x at
time t in the layer i satisfy the following set of conserva-
tion equations :

$$\frac{\partial}{\partial t} P(x,i,t) = k_i^+ P(x,i-1,t) + k_i^- P(x,i+1),t)$$

$$- (k_i^+ + k_i^-)P(x,i,t) - \frac{\partial}{\partial x} [u_i P(x,i,t)] \qquad (25)$$

Before proceeding, we note that this model has applications
in problems such diverse as petro-engineering, chromatography,
Taylor diffusion, spin diffusion in one dimensional chains,
residence time problems in reactors, phase diffusion in limit
cycles, etc. (see C. Van den Broeck and R.M. Mazo (1984)).
    In analogy with eq. (4), the displacement of the suspen-
ded particle at time t, x(t), is the sum of a large number
of displacements Δx_i during subsequent time intervals
[(i-1) t,iΔt], i = 1,...,N (NΔt = t). Moreover, as the particle "wanders" up and down the layers, it forgets about its
earlier positions. Hence, if Δt is choosen larger than the
characteristic memory time, then the displacements Δx_i are
(almost) independent one of another and we conclude (cf. eq.
(5)) :

$$<\delta x^2 (t)> \sim t \qquad (26)$$

The proportionality constant, called the effective longitudinal

diffusion coefficient K has been calculated exactly for the
above model (C. Van den Broeck and R.M. Mazo (1983) :

$$\lim_{t \to +\infty} \frac{<\delta x^2(t)>}{2t} = K = \sum_{r=1}^{n-1} \frac{[\sum_{i=1}^{n} \sum_{j=1}^{r} (u_i - u_j) \, P_i^{st} P_j^{st}]^2}{k_r^+ \, P_r^{st}} \qquad (27)$$

with

$$P_r^{st} = N \, k_1^+ k_2^+ \cdots k_{r-1}^+ k_{r+1}^- \cdots k_{n-1}^- k_n^- \qquad (28)$$

and N is a normalisation constant

$$(\sum_{r=1}^{n} P_r^{st} = 1)$$

It must be stressed that the result (26) heavily rests upon
the sufficiently fast decay of the time correlation function.
This is best illustrated by considering the continuum limit
of the above model in the case of a linear velocity field.
The vertical direction (perpendicular to the layers) will be
denoted by y, and the velocity field reads u(y) = αy. The
probability density P(x,y,t) obeys the following Fokker
Planck equation

$$\frac{\partial}{\partial t} P(x,y,t) = D \frac{\partial^2}{\partial y^2} P(x,y,t) - \frac{\partial}{\partial x} [u(y)P(x,y,t)] \qquad (29)$$

A straightforward calculation yields :

$$<\delta x^2(t)> = \frac{2}{3} \alpha^2 Dt^3 + <\delta x^2(t=0)> \qquad (30)$$

hence

$$\lim_{t \to +\infty} \frac{<\delta x^2(t)>}{t} = + \infty$$

The reason for the breakdown of the result (26) is the existence of an infinite correlation time. A particle never forgets
its initial position, as it is shown by the exact result for
the time correlation function :

$$<\delta x(t)\delta x(t')> = <\delta x^2 (\min(t,t')>$$ (31)

C. Van den Broeck
Dept. Natuurkunde
Vrije Universiteit Brussel
Pleinlaan 2, 1050 Brussel, Belgium.

REFERENCES

B.J. Berne and R. Pecora (Eds.), Dynamic Light Scattering
    (Wiley, New York, 1976).
A. Einstein, Ann. Phys. 17, 549 (1905).
P. Kotelenez, Ph. Thesis, University of Bremen (1983).
T.G. Kurz, J. Chem. Phys. 57, 2976 (1972).
S.K. Ma, Modern Theory of Critical Phenomena (Benjamin, N.Y.,
    1975).
M. Malek Mansour, C. Van den Broeck, G. Nicolis and J.W.
    Turner, Ann. Phys. 131, 283 (1981).
C. Van den Broeck, M. Malek Mansour and J. Houard, Physica
    101A, 167 (1980).
C. Van den Broeck and R.M. Mazo, Phys. Rev. Lett. 51, 1309
    (1983).
C. Van den Broeck and R.M. Mazo, J. Chem. Phys. (to appear,
    1984).
N.G. Van Kampen, Can. J. Phys. 39, 551 (1961).

B. Grigelionis, R. Mikulevičius

ON THE FUNCTIONAL LIMIT THEOREMS

Introduction

In the recent years an increasing interest is paid
to the functional limit theorems for infinite
dimensional stochastic processes, related to the
investigations of stochastic space-time models in
physics, chemistry, biology and other fields as
well as in the general theory of stochastic
evolution equations (see, e.g. [1] - [20]). Most
of the known results are concerned with the cases
of random processes taking values in Hilbert spaces,
spaces of measures and nuclear spaces, in particular,
spaces of distributions. Note that sometimes it is
necessary to consider a weak convergence of
probability measures on nonmetrisable topological
spaces. For example, in the case of weak convergence
to the solutions of stochastic evolution equations
there are considered semimartingales
$X_t = X_0 + A_t + Z_t$, $t \geq 0$, taking values in the
rigged Hilbert space $V \subset H \equiv H' \subset V'$, where V
is a separable reflexive Banach space, which is
dense in H, moreover the imbeddings $V \to H$ and
$H' \to V'$ are continuous, A_t, $t \geq 0$, is a V'-valued
stochastic process with the finite variation on each
finite time interval, Z_t, $t \geq 0$, is a H-valued

L. Arnold and P. Kotelenez (eds.), Stochastic Space-Time Models and Limit Theorems, 191–215.
© *1985 by D. Reidel Publishing Company.*

semimartingale, satisfying the condition

$\int_0^t |X_s|_V |dA_s|_{V'} < \infty$ for each $t > 0$.   Under some

natural assumptions it reduces to the investigation
of stochastic processes with the paths in the
function space $X = D_{[0,\infty)}(H) \cap \cap L_{loc}^p(V,N)$,   where
$D_{[0,\infty)}(H)$ is the Skorokhod space,

$L_{loc}^p(V,N) = \{\hat{w}(\cdot): \int_0^t |\hat{w}(s)|_V^p dN_s < \infty, \forall t \geq 0\}$,

$p \geq 1$,   and N is an increasing continuous function.
The space X is considered to be equipped with the
supremum of J_1-topology and the topology of weak
convergence in $L_{loc}^p(V,N)$.   It is important to note,
that such topological spaces are completely regular
and the criterion of Prokhorov for weak relative
compactness of the sets of probability measures on
them is still valid (see [21]).

It is well known that in the functional limit
theorems for finite dimensional as well as infinite
dimensional stochastic processes the method based on
the martingale characterization of limiting points,
combined with the known criteria of relative
compactness and the laws of large numbers, is rather
effective and universal. Following [22], the aim of
this paper is the formalization of this method for
arbitrary topological spaces and its applications
to the weak convergence of semimartingales taking
values in the rigged Hilbert spaces.

1. Martingale characterization and weak convergence of probability measures on topological spaces

Let X be an arbitrary topological space, $B(X)$ be the σ-algebra of the Borel subsets, $B = \{B_t, t \geq 0\}$ be some increasing family of sub-σ-algebras, $B_t \subset B(X)$, $t \geq 0$,

$B_+ = \{B_{t+}, t \geq 0\}$, $B_{t+} = \bigcap_{\varepsilon > 0} B_{t+\varepsilon}$, $t \geq 0$, $\hat{P}$ be a probability measure on $B(X)$, $\{M_t^i(x), t \geq 0, x \in X, i \in I\}$ a family of functionals being right continuous in t with left limits, where I is some set. We shall assume that the measure $\hat{P}$ is uniquelly defined by the assumptions that $\hat{P}|_{B_0} = \mu_0$ is given and for each $i \in I$ M^i are $(B_+, \hat{P})$-local martingales (for terminology see, e.g., [23] - [25]).

Let us consider further a sequence $\{X_n, n \geq 1\}$ of random elements, defined on probability spaces (Ω_n, F_n, P_n) with the given filtrations of σ-algebras $F_n = \{F_{nt}, t \geq 0\}$ and taking values in X, $n \geq 1$. Denote $\hat{P}_n = P_n \circ X_n^{-1}$, $\hat{P}_n|_{B_0} = \mu_0^{(n)}$, $n \geq 1$. We shall assume that we are given the families $\{M^{(n),i}, i \in I\}$ of (F_n, P_n)-local martingales, in terms of which the conditions of weak convergence $\hat{P}_n \Rightarrow \hat{P}$ as $n \to \infty$ will be formulated.

Denote by $T(F_n)$ the class of stopping times (s.t.) with respect to the filtration F_n, and for each $T > 0$ by $\Sigma(T)$ the class of such sequences $\{S_n, n \geq 1\}$ of s.t. $S_n = S_n(T) \in T(F_n)$, that $S_n \leq T$ and

$$\lim_{n \to \infty} P_n\{S_n < T\} = 0 . \tag{1}$$

Let M_X be a class of limiting points of the sequence $\{\hat{P}_n, n \geq 1\}$ and K be a class of increasing functions $G: R_+ \to R_+$ such that $G(t)/t \to \infty$ as $t \to \infty$.

Assume that for each $\nu \in M_X$ and $i \in I$ there exists a denumerable dense subset $Q = Q^{\nu,i} \subset R_+$ and $G = G^{\nu,i} \in K$, such that the following conditions are satisfied:

(α) $B_t = \sigma\{f: f\in B_t\}$ for each $t \in Q$, where B_t is some set of ν - a.e. continuous bounded functions on X;

(β) for each $t \in Q$ and $i \in I$ the functional $M_t^i(\cdot)$ is ν - a.e. continuous;

(γ) for some sequence $\{T_k, k \geq 1\}$, $o \leq T_k \uparrow \infty$ there exist sequences $\{S_n^{(k)}, n \geq 1\} \in \Sigma(T_k)$, such that for each $t \in Q$, $i \in I$ and $k \geq 1$

$$\sup_{n \geq 1} E_n \, G(|M_{t \wedge S_n^{(k)}}^{(n),i}|) < \infty \tag{2}$$

The following assertion holds.

Theorem 1. Assume that:

1) the sequence $\{\hat{P}_n, n \geq 1\}$ is weakly relative compact;

2) for some subclass of sets $B \in B_o$, defining μ_o,

$$\lim_{n \to \infty} \mu_o^{(n)}(B) = \mu_o(B);$$

3) the assumptions (α) - (γ) are satisfied;

4) for each $\epsilon > 0$, $t \in \bigcup_{\nu \in M_X} Q^{i,\nu}$, $i \in I$

$$\lim_{n\to\infty} P_n\{|M_t^{(n),i}-M_t^i(X_n)| > \varepsilon\} = 0 .$$

Then $\hat{P}_n \Rightarrow \hat{P}$ as $n \to \infty$.

P r o o f of the theorem 1 follows directly from the results of § 1 of the paper [22].

2. Canonical form of semimartingales with values in rigged Hilbert spaces

Stochastic processes, considered in this paragraph, are assumed to be defined on a probability space (Ω,F,P) and adapted to a filtration of σ-algebras $F = \{F_t, t\geq 0\}$.

For a normed vector space E we shall denote by $V(E)$ a class of right continuous E-valued processes with left limits having P - a.e. finite variation on each finite time interval, $A(E)$ a subclass of processes of $V(E)$ which have summable variation, $M(E)$ a class of uniformly integrable E-valued (P,F)-martingales, $M^2(E)$ a subclass of $M(E)$ of square integrable martingales, $S(E)$ a class of E-valued (P,F)-semimartingales, $S_p(E)$ a class of E-valued (P,F)-special semimartingales. As usual for each class of processes C we denote by C^c a subclass of continuous elements of C and by C_{loc} a class of processes $C = \{C_t, t\geq 0\}$, for which there exists a sequence $\{T_n, n\geq 1\}$, $T_n \in T(F_n)$, $T_n \uparrow \infty$, such that $\{C_{t\wedge T_n}-C_o, t\geq 0\} \in C$ for all $n \geq 1$. P will denote a σ-algebra of F-predictable subsets of $[0,\infty) \times \Omega$. For two normed vector spaces $E_1 \supseteq E_2$ and $e \in E_1$ we shall denote by $|e|_{E_2}$

the norm of e in E_2, if $e \in E_2$, and $= \infty$ if $e \notin E_2$.

Let us consider a rigged Hilbert space $V \subset H \equiv H' \subset V'$ and a stochastic process $X = \{X_t = A_t + Z_t, t \geq 0\}$, where $A \in V(V')$, $A_0 = 0$, $Z \in S(H)$ and for each $t > 0$ P a.e. $\int_0^t |X_s|_V d|A|_s < \infty$; $|A|_t = \int_0^t |dA_s|_V$, . Denote by $p(du,dx)_\Pi$ the jump measure of X, $\Pi(du,dx)$ the dual predictable projection (the compensator) of $p(du,dx)$, $q(du,dx) = p(du,dx) - \Pi(du,dx)$.

Let $h: H \rightarrow H$ be a continuous bounded function, $h(x) = x$ for $|x|_H \leq \frac{1}{2}$ and $= 0$ for $|x|_H \quad 1$.

Theorem 2. The paths of the process X P - a.e. are H-valued right continuous with left limits in the strong topology of H . Moreover

$$|X_t|_H^2 = |X_0|_H^2 + 2\int_0^t X_s dA_s + 2\int_0^t A_{s-} dZ_s +$$

$$+ |Z_t|_H^2 - \sum_{s \leq t} |\Delta A_s|_H^2, \quad t \geq 0, \qquad (4)$$

and the following unique decompostion holds:

$$X_t = X_0 + \alpha_t^h + X_t^c + \int_0^t \int_{0 \Pi \setminus \{o\}} h(x) q(du,dx) + \int_0^t \int_{0 H \setminus \{o\}} (x - h(x)) p(du,dx),$$
(5)

where $\alpha^h \in A_{loc}(V') \cap P$, $X^c \in M_{loc}^c(H)$, $\Delta\alpha_t^h = t \geq 0$,

$= \int_{H \setminus \{o\}} h(x) \Pi(\{t\} \times dx)$, $t \geq 0$, and there exist $C \in V(V')$, $S \in S(H)$, such that P - a.e. for each $t > 0$ $\int_0^t |X_s|_V d|C|_s < \infty$ and $\alpha^h = C + S$.

P r o o f . The first part of theorem 2 is proved mainly following the papers [26], [27], containing more special results.

Let $Z = D + M$, $D \in V(H)$, $M \in M_{loc}(H)$,

$$\beta_t = \inf\{s: |A|_s > t\}, \quad |D|_t = \int_0^t |dD_s|_H, \quad t \geq 0.$$

Without restriction of generality we shall assume, that

$$\beta_0 = 0, |A|_\infty + |D|_\infty + |X_0|_H^2 + \int_0^\infty |X_s|_V d|A|_s + \sup_{t \geq 0} |M_t|_H \leq 1.$$

We shall need later several lemmas. By means of the standard change of time we obtain the following assertion.

Lemma 1. If g is $F \otimes B(\mathbb{R}^1)$-measurable function and $\int_0^t |g(u)| d|A|_u < \infty$, then

$$\int_0^t g(u) d|A|_u = \int_0^{|A|_t} g(\beta_s) ds .$$

In addition, $|A|_{\beta_t-} - |A|_{\beta_s} \leq t - s, \quad s < t$.

Denote

$$\psi_n^{(1)}(s) = \frac{i}{2^n}, \quad s \in [\frac{i}{2^n}, \frac{i+1}{2^n}), \quad i = 0, \pm 1, \ldots,$$

$$\psi_n^{(2)}(s) = \frac{i+1}{2^n}, \quad s \in (\frac{i}{2^n}, \frac{i+1}{2^n}], \quad i = 0, \pm 1, \ldots$$

Lemma 2. Let $\{g_k, k \geq 1\}$ be a sequence of V-valued $B \otimes B(\mathbb{R}^1)$-measurable functions such that $g_k(s) = 0$ for $s \notin [0,1]$ and $E[\int_0^1 |g_k(s)|_V ds] < \infty, k \geq 1$. Then there exists a sequence $n_j \uparrow \infty$ such that for almost all $u \in [0,1]$ with respect to the Lebesgue measure

$$\lim_{j\to\infty} E[\int_0^1 |g_k(s) - g_k(\psi_{n_j}^{(i)}(s-u)+u)|_v ds] = 0, \quad k \geq 1, \quad i=1,2.$$

P r o o f . For each $k \geq 1$ and $\varepsilon > 0$ there exists a $F \otimes B(\mathbb{R}^1)$-measurable function $g_{k,\varepsilon}(s)$, continuous in s , such that

$$E[\int_{-\infty}^{\infty} |g_k(s) - g_{k,\varepsilon}(s)|_v ds] \leq \varepsilon$$

It follows that for each $t \in \mathbb{R}^1$, $k \geq 1$

$$\lim_{h\to 0} [\int_{\infty}^{\infty} |g_k(s+t+h) - g_k(s+t)|_v ds] = 0 .$$

and therefore as $n \to \infty$

$$n_{k,n}^{(i)}(t) = E[\int_0^{\infty} |g_k(s+\psi_n^{(i)}(t)) - g_k(s+t)|_v ds] \to 0 ,$$

moreover $n_{k,n}^{(i)}(t) = 0$ for $t > 2$.

Thus as $n \to \infty$

$$\int_0^{\infty} \int_{-1}^{\infty} E|g_k(s+\psi_n^{(i)}(t)) - g_k(s+t)|_v ds\, dt =$$

$$= \int_0^{\infty} \int_{t-1}^{\infty} E|g_k(s+\psi_n^{(i)}(s-t)) - g_k(s)|_v ds\, dt \to 0$$

and we can find $n_k \uparrow \infty$ such that

$$\sum_{i=1}^{2} \sum_{j=1}^{k} \int_0^{\infty} \int_{t-1}^{\infty} E|g_j(t+\psi_n^{(i)}(s-t)) - g_j(s)|_v ds\, dt \leq 2^{-k}$$

for each $n \geq n_k$. So we have that for almost all $t \in [0,1]$ and $k \geq 1$

$$\sum_{i=1}^{2} \sum_{j=1}^{\infty} \int_0^{\infty} E|g_k(t+\psi_{n_j}^{(i)}(s-t)) - g_k(s)|_v ds < \infty .$$

Lemma 2 is proved.

Let $\{e_j, j \geq 1\}$ be a basis of H consisting elements from V, $z_t^{(k)} = \sum_{j=1}^{k} (Z_t, e_j)_H e_j$. Further we agree for vector-valued function f, defined on $[0, \infty)$, to assume that $f(\infty) = 0$.

Using lemma 1 we have that for each $k \geq 1$

$$E[\int_0^1 (|X_{\beta_s}|_V + |Z_{\beta_s}^{(k)}|_V) ds < \infty .$$

Thus from lemma 2 we deduce the following assertion.

Corollary 1. There exists a sequence of the imbedded partitions $0 = s_0^{(n)} < s_1^{(n)} < \ldots < s_{l_n}^n = 1$, $\max_{1 \leq j \leq l_n} (s_{j+1}^{(n)} - s_j^{(n)}) \to 0$, as $n \to \infty$, such that for all $k \geq 1$, $i = 1, 2$

$$\lim_{n \to \infty} E[\int_0^\infty |X_{\beta_s} - \bar{X}_{\beta_{\varphi_n^{(i)}(s)}}|_V ds] = 0$$

and

$$\lim_{n \to \infty} E[\int_0^\infty |Z_{\beta_s}^{(k)} - \bar{Z}_{\beta_{\varphi_n^{(i)}(s)}}^{(k)}|_V ds] = 0 ,$$

where

$$\varphi_n^{(1)}(s) = s_j^{(n)}, \quad s \in [s_j^{(n)}, s_{j+1}^{(n)}), \quad j = 1, \ldots, l_n ,$$

$$\varphi_n^{(2)}(s) = s_{j+1}^{(n)}, \quad s \in (s_j^{(n)}, s_{j+1}^{(n)}], \quad j = 1, \ldots, l_n ,$$

$$\bar{X}_0 = \bar{Z}_0^{(k)} = 0, \quad \bar{X}_s = X_s, \quad \bar{Z}_s^{(k)} = Z_s^{(k)} \quad \text{for} \quad s > 0.$$

Let further $\tau_j^{(n)} = \beta_{s_j}(n)$, $j = 1, \ldots, l_n$, $X_n^{(1)}(s) = 0$ for $0 \leq s < \tau_1^{(n)}$, $= X_{\tau_j}(n)$, for $s \in [\tau_j^{(n)}, \tau_{j+1}^{(n)})$, $j = 1, \ldots, l_n$, $X_n^{(2)}(s) = X_{\tau_1}(n)$ for $0 < s \leq \tau_1^{(n)}$,

$= X_{\tau_{j+1}^{(n)}}(n)$ for $s \in (\tau_j^{(n)}, \tau_{j+1}^{(n)}]$, $j = 1, \ldots, l_n - 1$.

The processes $z_n^{(k),i}$, $i = 1, 2$, are defined in the analogous way.

Corollary 2. The following equalities hold:

$$\lim_{n \to \infty} E[\int_0^\infty |X_u - X_n^{(i)}(u)|_V d|A|_u] = 0, \quad i = 1, 2,$$

$$\lim_{n \to \infty} E[\int_0^\infty |z_u^{(k)} - z^{(k),i}(u)|_V d|A|_u] = 0, \quad i = 1, 2, \quad k \geq 1.$$

Indeed, from lemma 1 and corollary 1 it follows that

$$\lim_{n \to \infty} E[\int_0^\infty |X_u - X_n^{(i)}(u)|_V d|A|_u] = \lim_{n \to \infty} E[\int_0^\infty |X_{\beta_s}$$

$$- X_n^{(i)}(\beta_s)|_V ds] \leq \lim_{n \to \infty} \sum_{i=1}^2 E[\int_0^\infty |X_{\beta_s} - \overline{X}_{\beta_{\varphi_n^{(i)}(s)}}|_V ds] = 0.$$

The second equality is proved similarly.

Corollary 3. We have, that

$$E[\sup_{u,n} |X_n^{(i)}(u)|_V] < \infty, \quad i = 1, 2.$$

It follows from the corollary 2 and the step form of $X_n^{(i)}$. It is easy to check that

$$|X_{\tau_{k+1}^{(n)}}(n)|_H^2 = |X_0|_H^2 + 2\int_0^{\tau_{k+1}^{(n)}} X_n^{(2)}(s) dA_s + 2\sum_{j=0}^k X_{\tau_j}(n)(z_{\tau_{j+1}}(n)$$

$$- z_{\tau_j}(n)) + \sum_{j=0}^k (|z_{\tau_{j+1}}(n) - z_{\tau_j}(n)|_H^2 - |A_{\tau_{j+1}}(n) - A_{\tau_j}(n)|_H^2) =$$

$$|X_0|_H^2 + 2\int_0^{\tau_{k+1}^{(n)}} X_n^{(2)}(s) dA_s + 2\sum_{j=0}^k (X_{\tau_j}(n) - z_{\tau_j}(n))(z_{\tau_{j+1}}(n) -$$

$$- Z_{\tau_j}(n)) + |Z_{\tau_{k+1}}(n)|_H^2 - \sum_{j=0}^{k} |A_{\tau_{j+1}}(n) - A_{\tau_j}(n)|_H^2 . \tag{6}$$

Using the inequality $2ab \le \varepsilon^2 a^2 + \frac{b^2}{\varepsilon^2}$, $\varepsilon > 0$, from (6) and corollary 2 we have that

$$E[\sup_{n,k} |X_{\tau_k}(n)|_H^2] < \infty \tag{7}$$

and then

$$E[\sup_{n,k} |A_{\tau_k}(n)|_H^2] < \infty . \tag{8}$$

Lemma 3.   P - a.e. for each $t > 0$ $A_t \in H, A$ is right continuous with left limits in the weak topology of H and

$$E[\sup_{t \ge 0}(|A_t|_H^2 + |A_{t-}|_H^2)] < \infty \tag{9}$$

P r o o f .   Set $I = I(w) = \bigcup_{n,k} \{\tau_k^{(n)}(w)\} \cap [0,\infty)$.
Remark that the closure $\bar{I}$ is the support of the measure $d|A|_t$ on $[0,\infty)$. So far as according to (8) P - a.e. $\sup_{t \in I} |A_t|_H^2 < \infty$, the set $\{A_t, t \in I\}$ is a weakly compact subset of H.   If $t_n \in I$, $t_n > t$ and $t_n \downarrow t$ (respectively, $t_n < t$, $t_n \uparrow t$), then $A_{t_n} \to A_t$ (respectively, $A_{t_n} \to A_{t-}$) and $|A_t|_H \le \sup_n |A_{t_n}|_H$ (respectively, $|A_{t-}|_H \le \sup_n |A_{t_n}|_H$). It is clear that $\bar{I}^C = [0,\infty) \setminus \bar{I}$ is the union of a finite or denumerable number of open intervals and A is constant on each of them. Lemma 3 is proved.

Lemma 4. Let $\{a_n, n \ge 1\}$ be a sequence of H-valued predictable processes such that $P\{\sup_{n,t} |a_n(t)|_H < \infty\} = 1$ and for each $t \ge 0$ P - a.e. $\lim_{n \to \infty} a_n(t) = 0$ in the weak topology of H. Then for each $\varepsilon > 0$

$$\lim_n P\{\sup_t |\int_o^t a_n(s)dZ_s| > \varepsilon\} = 0 .$$

P r o o f . Denote π^k the projector to the subspace, generated by the k first elements of a basis in H. Then

$$P\{\sup_t |\int_o^t a_n(s)dZ_s| > \varepsilon\} \le P\{\sup_t |\int_o^t a_n(s)dD_s| > \tfrac{\varepsilon}{2}\} +$$

$$+ P\{\sup_t |\int_o^t a_n(s)dM_s| > \tfrac{\varepsilon}{2}\} \tag{10}$$

and for each $k \ge 1$ as $n \to \infty$

$$P\{\sup_t |\int_o^t a_n(s)dD_s| > \tfrac{\varepsilon}{2}\} \le P\{\sup_t |\int_o^t \pi^k a_n(s)dD_s| > \tfrac{\varepsilon}{4}\} +$$

$$+ P\{[\sup_t |a_n(t)|_H |(I-\pi^k)D|_\infty] > \tfrac{\varepsilon}{2}\} \to 0 \tag{11}$$

according to the assumptions of the lemma and the fact that $|I-\pi^k)B|_\infty \to 0$ in probability as $k \to \infty$.

Using the known inequality (see [28]) for each $\delta > 0$ and $\varepsilon > 0$

$$P\{\sup_t |\int_o^t a_n(s)dM_s| > \tfrac{\varepsilon}{2}\} \le P\{\sup_t |\int_o^t \pi^k a_n(s)dM_s| > \tfrac{\varepsilon}{4}\} +$$

$$+ P\{\sup_t |\int_o^t a_n(s)d[(I-\pi^k)M_s]| > \tfrac{\varepsilon}{4}\} \le$$

$$\le P\{\int_o^\infty |\pi^k a_n(s)|_H^2 d<M>_s > \delta\} + \frac{16\delta}{\varepsilon^2} +$$

$$+ P\{\int_o^\infty |a_n(s)|_H^2 d<(I-\pi^k)M>_s > \delta\} + \frac{16\delta}{\varepsilon^2} +$$

$$\le P\{\int_o^\infty |\pi^k a_n(s)|_H^2 d<M>_s > \delta\} + \frac{32\delta}{\varepsilon^2} +$$

$$+ P\{[\sup_t |a_n(t)|_H < (I-\pi^k)M>_\infty] > \delta\}. \tag{12}$$

From (11)-(12) it follows the statment of the lemma 4.

Define

$$A_n(u) = A_{\tau_j(n)}, \quad u \in [\tau_j^{(n)}, \tau_{j+1}^{(n)}), \quad j = 0,1,\ldots,l_n-1.$$

Lemma 5. We have that for each $\varepsilon > 0$

$$\lim_{n\to\infty} P\{\sup_t |\int_o^t A_n(u)dz_u - \int_o^t A_{u-}dzu| > \varepsilon\} = 0 .$$

This statement follows from the lemma 3 and the fact that for each $u > 0$ $A_n(u) - A_{u-} \to 0$ weakly in H as $n \to \infty$.

Denote $U = \{\tau_j^{(n)}, 0\leq j\leq l_n, n\geq 1$,

$$K_n(t) = \sum_{\tau_{j+1}^{(n)} \leq t} |A_{\tau_{j+1}(n)} - A_{\tau_j(n)}|_H^2 ,$$

$$K_t = \sum_{s\leq t} |\Delta A_s|_H .$$

Lemma 6. For $t \in U$ in probability

$$\lim_{n\to\infty} K_n(t) = K_t.$$

P r o o f . From the equality (6) it follows the existence of the limit in probability $\lim_{n\to\infty} K_n(t) = \overline{K}_t.$ So as far

$$\sum_{\tau_{j+1}^{(n)} \leq t} |A_{\tau_{j+1}(n)} - A_{\tau_j(n)}|_H^2 = \sum_{k=1}^{\infty} \sum_{\tau_{j+1}^{(n)} \leq t} |(A_{\tau_{j+1}(n)} - A_{\tau_j(n)}, e_k)_H|^2 \tag{13}$$

and

$$\lim_{n\to\infty} \sum_{\tau_{j+1}^{(n)} \leq t} |(A_{\tau_{j+1}(n)} - A_{\tau_j(n)}, e_k)_H|^2 = \sum_{s\leq t} |(\Delta A_s, e_k)_H|^2, \tag{14}$$

then from (13)-(14) and the Fatou lemma we conclude
that   P - a.e. $K_t \leq \bar{K}_t$, $t \geq 0$.

Now we have to prove the reverse inequality. We
have

$$\sum_{\tau_{j+1}^{(n)} \leq t} |A_{\tau_{j+1}(n)} - A_{\tau_j(n)}|_H^2 = \sum_{\tau_{j+1}^{(n)} \leq t} \{|\Delta A_{\tau_{j+1}(n)}|_H^2 - |A_{\tau_{j+1}(n)} - A_{\tau_j(n)}|_H^2 +$$

(15)

$$+ 2[(A_{\tau_{j+1}(n)} - A_{\tau_j(n)}, X_{\tau_{j+1}(n)} - X_{\tau_j(n)})_H +$$

$$(A_{\tau_{j+1}^n} - A_{\tau_j(n)}, Z_{\tau_{j+1}(n)} - Z_{\tau_j(n)})_H]\}.$$

But as $n \to \infty$ P - a.e.

$$\sum_{\tau_{j+1}^{(n)} \leq t} |(A_{\tau_{j+1}(n)} - A_{\tau_j(n)}, X_{\tau_{j+1}(n)} - X_{\tau_j(n)})_H| \leq \int_0^\infty |X_n^{(1)}(s) -$$

$$- X_n^{(2)}(s)|_V d|A|_s + |(X_0, A_{\tau_1(n)_-})_H| \to 0.$$

(16)

Further according to the corollary 2 and the
assumption that $Z \in S(H)$, as $n \to \infty$ P - a.e.

$$\sum_{\tau_{j+1}^{(n)} \leq t} [2(A_{\tau_{j+1}(n)} - A_{\tau_j(n)}, Z_{\tau_{j+1}(n)} - Z_{\tau_j(n)})_H - |A_{\tau_{j+1}(n)} - A_{\tau_j(n)}|_H^2]$$

$$\leq \sum_{\tau_{j+1}^{(n)} \leq t} [|(Z_{\tau_{j+1}(n)} - Z_{\tau_{j+1}(n)}^{(k)}) - (Z_{\tau_j(n)} - Z_{\tau_j(n)}^{(k)})|_H^2 +$$

$$+ 2(A_{\tau_{j+1}(n)} - A_{\tau_j(n)}, Z_{\tau_{j+1}(n)}^{(k)} - Z_{\tau_j(n)}^{(k)})_H|] \to 0.$$

(17)

Thus from (15) - (17) we obtain that P - a.e.
$\bar{K}_t \leq K_t$, $t \geq 0$.   Lemma 6 is proved.

From formula (6) and the lemmas proved above we have that P - a.e. for each $t \in I$

$$|X_t|^2_H = |X_0|^2_H + 2 \int_0^t X_s dA_s + 2 \int_0^t A_{s-} dz_s + |z_t|^2_H -$$

$$- \sum_{s \leq t} |\Delta A_s|_H \cdot \tag{18}$$

P - a.e., if $t \in \bar{I} \smallsetminus I$, and
$\sigma^{(n)} = \inf\{\tau^{(n)}_{j+1} : \tau^{(n)}_j \leq t < \tau^{(n)}_{j+1}\} \downarrow t$, then

$$|X_\sigma(n) - X_\sigma(n+p)|^2_H = |X_\sigma(n+p)|^2_H - |X_\sigma(n)|^2_H -$$

$$- 2(X_\sigma(n), X_\sigma(n+p) - X_\sigma(n))_H \to 0,$$

because according to (18) as $n \to \infty$ P - a.e.

$$|X_\sigma(n+p)|^2_H - |X_\sigma(n)|^2_H \to 0$$

and

$$|(X_\sigma(n), X_\sigma(n+p) - X_\sigma(n))_H| \leq |(X_\sigma(n), z_\sigma(n+p) - z_\sigma(n))_H| +$$

$$+ |\int_{\sigma^{(n)}}^{\sigma^{(n+p)}} (X_n^{(2)}(s) - X_s) dA_s| + |\int_{\sigma^{(n)}}^{\sigma^{(n+p)}} X_s dA_s| \leq \text{const } z_\sigma(n+p)$$

$$-z_\sigma(n)|_H + \int_0^\infty |X_2^{(2)}(s) - X_s| \, d|A|_s + \int_{\sigma^{(n)}}^{\sigma^{(n+p)}} |X_s|_v d|A|_s \to 0.$$

Thus P - a.e. $X_\sigma(n) \to X_t$ in the strong topology of H .

Analogously P - a.e. for $t \in \bar{I} \smallsetminus I$ and
$\hat{\sigma}^{(n)} = \max\{\tau^{(n)}_j : \tau^{(n)}_j < t \leq \tau^{(n)}_{j+1}\} \uparrow t$ we have that
$X_\sigma(n) \to X_{t^-}$ in the strong topology of H. If
$t, s \in \bar{I}^C$ and are contained in the same open

interval, then $|X_t - X_s|_H^2 = |Z_t - Z_s|_H^2$. From what was said above it follows that X P - a.e. has the right continous paths with left limits. Together with the formula (18) it proves the formula (4).

Now we shall prove the formula (5). We have

$$X_t = X_o + F_t + G_t + \int_o^t \int_{H \setminus \{o\}} (x - h(x)) p(du, dx), \quad t \geq 0,$$

where

$$F_t = A_t + \sum_{s \leq t} \Delta Z_s X\{|\Delta Z_s|_H > 1\} \int_o^t \int_{II \setminus \{o\}} (x - h(x)) p(du, dx)$$

and

$$G_t = Z_t - \sum_{s \leq t} \Delta Z_s X\{|\Delta Z_s|_H > 1\} .$$

It is clear that $\sup_t |\Delta F_t|_H \leq const$, $F \in A_{loc}(V')$, $G \in Sp(H)$. The following unique decompositions hold (see, e.g. [24]):

$$F_t = \tilde{F}_t + L_t, \quad G_t = \tilde{G}_t + R_t, \quad t \geq 0,$$

where $F \in A_{loc}(V') \cap P$, $L \in M_{loc}(V')$,
$\tilde{G} \in A_{loc}(H) \cap P$, $R \in M_{loc}^2(H)$.

According to the formula (4) P - a.e. $\sum_{s \leq t} |\Delta X_s|_H^2 < \infty, t \geq 0$

Therefore the integral $\int_o^t \int_{H \setminus \{o\}} h(x) q(du, dx)$, $t \geq 0$, is defined and $X^c \in M_{loc}^c(V')$, where

$$X_t^c = X_t - X_o - \int_o^t \int_{H \setminus \{o\}} (x - h(x)) p(du, dx) - \int_o^t \int_{H \setminus \{o\}} h(x) q(ds, dx) -$$

$$- \tilde{F}_t - \tilde{G}_t, \quad t \geq 0.$$

We have that

$$X_t^c + \int_0^t \int_{H\setminus\{o\}} h(x)q(du,dx) = L_t + R_t^c + R_t^d, \quad t \geq 0,$$

where R^c and R^d are continuous and purely discontinuous parts of R, respectively. For each $v \in V$ we shall have

$$<X_t^c,v> + \int_0^t \int_{H\setminus\{o\}} (h(x),v)_H \, q(du,dx) = <L_t,v> +$$

$$+ (R_t^c,v)_H + (R_t^d,v)_H, \quad t \geq 0. \tag{20}$$

As far as $<L_t,v>$, $t \geq 0$, is a purely discontinuous local martingale, then from (19) - (20) we find that

$$L_t + R_t^d = \int_0^t \int_{H\setminus\{o\}} h(x)q(du,dx), \quad t \geq 0,$$

and $R_t^c = X_t^c$, $t \geq 0$, i.e. $L \in M_{loc}^2(H)$. The formula (5) now follows from (19), if we denote $a_t^h = \tilde{F}_t + \tilde{G}_t$, $t \geq 0$. Theorem 2 is proved.

Remark 1. Under the assumptions of theorem 2 we can correctly define the integral $\int_0^t X_{s-}da_s^h$, $t \geq 0$, by means of the following formula:

$$\int_0^t X_{s-}da_s^h = \int_0^t X_s dC_s - \sum_{s\leq t} (\Delta X_s, \Delta C_s)_H + \int_0^t X_{s-}dS_s, t \geq 0.$$

Remark 2. The triplet (a^h,β,Π) where β is defined by the equality

$$<(X^c,v)_H>_t = (v,\beta_t v)_H, \quad t \geq 0, v \in V,$$

is called the triplet of the $(P,\mathbb{F})$-predictable

characteristics of the process X . The class of
processes X , permitting the decomposition (5), we
shall denote by $S(V,H,P,\mathbb{F})$. Note that for functions
h_1 and h_2, satisfying analogous assumptions to h,
we have the equality:

$$\alpha_t^{h_2} = \alpha_t^{h_1} + \int_0^t \int_{H\smallsetminus\{o\}} (h_1(x)-h_2(x))\Pi(ds,dx), \quad t \geq 0 .$$

3. Weak convergence of semimartingales with values in a rigged Hilbert space

Now we shall consider the space
$X = D_{[o,\infty)}(H) \cap L^2_{loc}(V,N)$ with the family of
σ-algebras $B_t = \sigma(X_s, s\leq t)$, $t \geq 0$, where
$X_t(\hat{\omega}) = \hat{\omega}(t)$, $\hat{\omega} \in X$, and the probability measure $\hat{P}$
on $B(X)$, assuming that it is uniquely defined by
the given restriction $\hat{P}|B_o = \mu_o$ and the assumption
that $X \in S(V,H,\hat{P},\mathbb{B}_+)$ with the given triplet
$(\hat{\alpha}^h,\hat{B},\hat{\Pi})$ of the $(\hat{P},\mathbb{B}_+)$-predictable characteristics
and $|d\hat{\alpha}_t^h|_{V'} \leq (1+|X_t|_V)dN_t$.

Denote
$$\hat{\Gamma}_v^h(t,\omega) = (v,B_tv)_H + \int_{H\smallsetminus\{o\}} (h(x),v)_H^2 \hat{\Pi}(\hat{\omega},[o,t]\times dx) -$$

$$- \sum_{s\leq t} (\int_{H\smallsetminus\{o\}} (h(x),v)_H \hat{\Pi}(\hat{\omega},\{s\}\times dx), \quad t\geq 0, \; v\in V,$$

$$M_t^\varphi = P_t(\varphi) - \hat{\Pi}_t(\varphi), \quad \varphi \in \hat{C}^+(H\smallsetminus\{o\}) ,$$

$$Y_t^h = X_t-X_o-\hat{\alpha}_t^h - \int_0^t \int_{H\smallsetminus\{o\}} (x-h(x))p(ds,dx), \quad t\geq 0;$$

$\hat{C}^+(H\smallsetminus\{o\})$ is a space of continuous bounded functions
which are equal to zero in the neighbourhood of zero,

$$\nu_t(\varphi) = \int_{H \smallsetminus \{o\}} \varphi(x) \nu([o,t] \times dx).$$

It is easy to check that $X \in S(V,H,\hat{P},\mathbb{B}_+)$ with the triplet $(\hat{\alpha}^h, \hat{B}, \hat{\Pi})$ of the $(\hat{P}, \mathbb{B}_+)$-predictable characteristics iff the families

$$\{(Y_t^h, v)_H, \ t \geq 0, \ v \in V\}, \ \{(Y_t^h, v)_H^2 - \hat{\Gamma}_v^h(t), \ t \geq 0, \ v \in V\}$$

and $M_t^\varphi, \ t \geq 0, \ \varphi \in \hat{C}^+(H \smallsetminus \{o\})\}$ are the families of the $(\hat{P}, \mathbb{B}_+)$-local martingales.

We shall assume that $X_n \in S(V,H,P_n,F_n)$ with the paths in the space X, the triplets $(\alpha^{h,(n)}, B^{(n)}, \Pi^{(n)})$ of the (P_n, F_n)-predictable characteristics and initial distributions $\mu_o^{(n)}$, $n \geq 1$.

An immediate consequence of the theorem 1 is the following theorem.

Theorem 3. Assume that:

1) the sequence $\{\hat{P}_n, \ n \geq 1\}$ is weakly relative compact;

2) $\mu_o^{(n)} \Rightarrow \mu_o$ as $n \to \infty$;

3) for each $t > 0$, $v \in V$ and $\varphi \in \hat{C}^+(H \smallsetminus \{o\})$ $(\hat{\alpha}_t^h(\cdot), v)_H$, $(v, \hat{B}_t(\cdot)v)_H$ and $\hat{\Pi}_t(\cdot, \varphi)$ are continuous in the topology of X;

4) for each $v \in V$, $\varphi \in \hat{C}^+(H \smallsetminus \{o\})$ and $T > 0$ there exist a sequence $\{S_n, \ n \geq 1\} \in \Sigma(T)$ and $G \in K$ such that

$$\sup_{n \geq 1} E_n[G(\Gamma_v^{h,(n)}(S_n) + \Pi_{S_n}^{(n)}(\varphi))] < \infty;$$

5) for each $\varepsilon > 0$, $t > 0$, $v \in V$ and $\varphi \in \hat{C}^+(H \smallsetminus \{o\})$

$$\lim_{n \to \infty} P_n\{|(\alpha_t^{h,(n)}, v)_H - (\hat{\alpha}_t^h(X_n), v)_H| > \varepsilon\} = 0,$$

$$\lim_{n\to\infty} P_n\{|\Gamma_v^{h,(n)}(t) - \hat{\Gamma}_v^h(t,X_n)| > \varepsilon\} = 0 ,$$

$$\lim_{n\to\infty} P_n\{|\Pi_t^{(n)}(\varphi) - \hat{\Pi}_t(X_n,\varphi)| > \varepsilon\} = 0 .$$

Then $\hat{P}_n \Rightarrow \hat{P}$ as $n \to \infty$.

At the end of the paper we shall give a sketch of the proof of a criterion for the weakly relative compactness of $\{\hat{P}_n, n \geq 1\}$.

Denote $\tilde{\Sigma}(T) = (T_{n,j})_{n,j\geq 1} : T_{n,j} \in T(\mathbb{F}_n)$,

$$T_{n,j} \leq T, \lim_{j\to\infty}\overline{\lim}_{n\to\infty} P_n\{T_{n,j} < T\} = 0\},$$

$$\bar{\Pi}^k = I - \Pi^k, k \geq 1, \bar{\Pi}^0 = I.$$

Theorem 4. Let the following assumptions hold:

1) $\{\mu_0^{(n)}, n\geq 1\}$ is weakly relative compact;

2) for each $T > 0$ and $K > 0$ there exist $(T_{n,j}) \in \tilde{\Sigma}(T)$, the basis $\{e_k, k\geq 1\}$ of H, consisting of the elements from V, $\lambda_{n,j} \in \Lambda^+ \cap P$ and $\Lambda_{n,k} \in A^+ \cap P$ such that for each $\varepsilon > 0$

$$\lim_{l\to\infty}\overline{\lim}_{n\to\infty} P_n\{\Pi^{(n)}([0,T]x\{x:|x|_H>l\} > \varepsilon \} = 0 ,$$

$$\sup_{n,j,k} |X_{n,j}(\infty) + A_{n,k,j}(\infty)| \leq \text{const},$$

$$\lim_{k\to\infty}\overline{\lim}_{n\to\infty} P_n\{A_{n,k,j}(\infty) > \varepsilon\} = 0$$

and on the set $]0,T_{n,j}]$

$$2<\bar{\Pi}^k X_n(s-), d\alpha_s^{(n)}>+2(\bar{\Pi}^k X_n(s-), \int_{1<|x|_H\leq K} x \Pi^{(n)}(ds,dx))_H^+$$

$$+ d<\bar{\Pi}^k X_n^c>_s + \int_{|x|_H\leq K} |\Pi^k x|_H^2 \Pi^{(n)}(ds,dx)+\varepsilon_k|\bar{\Pi}^k X_n(s-)|_V^2 dN_s$$

$$\leq \; |X_n(s-)|_H^2 dX_{n,j}(s) + dA_{n,k,j}(s),$$

where $\varepsilon_o > 0$, $\varepsilon_k \geq 0$, $k \geq 1$;

3) for each $T_n \in T(\mathbb{F}_n)$, $T_n \leq T$, $\delta_n \downarrow 0$, $j \geq 1$ and $\varepsilon > 0$

$$\lim_{n \to \infty} P_n\{| (\alpha^{(n)}_{T_n+\delta_n}, e_j)_H - (\alpha^{(n)}_{T_n}, e_j)_H| + \Gamma^{h,(n)}_{e_j} (T_n+\delta_n) -$$

$$- \Gamma^{h,(n)}_{e_j}(T_n) + \Pi^{(n)}((T_n,T_n+\delta_n] \times \{x: |x|_H > 1\}) > \varepsilon\} = 0.$$

Then $\{\hat{P}_n, n \geq 1\}$ is weakly relative compact.

P r o o f .   From assumption 2) we have that

$$\lim_{1 \to \infty} \overline{\lim_{n \to \infty}} \; P_n\{\max_{u \leq T} |\Delta X_n(u)|_H > 1\} = 0 \; .$$

Therefore, denoting $S_{n,1} = \inf\{t: |\Delta X_n(t)|_H > 1\}$ we shall have that $(S_{n,1}) \in \tilde{\Sigma}(T)$. From the definition of $\tilde{\Sigma}(T)$ and the general form of compact subsets in the spaces $D_{[0,\infty)}(H)$ and $L^2_{loc}(V,N)$ it is clear that it is enough to prove for each $T > 0$, $j \geq 1$ and $1 \geq 1$ the weakly relative compactness of the sequence of measures, corresponding to the processes $X_{n,T_{n,j}} \wedge S_{n,1}$ - (we shall denote them later by the same letter X_n), where $f_{S-}(t) = f(t)$ for $t < S$ and $= f(S-)$ for $t \geq S$. Besides, according to 1), we can assume that $\sup |X_n(o)|_H \leq$ const and $\lim_{k \to \infty} \sup_n E_n |\bar{\Pi}^k X_n(o)|_H = 0$.

Using the formula (4) we have that

$$|\bar{\Pi}^k X_n(t)|_H^2 = |\bar{\Pi}^k X_n(o)|_H^2 + 2 \int_o^t \bar{\Pi}^k X_n(s-) d\alpha_n(s) +$$

$$+ 2\int_o^t \int_{|x|_H>1} (\bar{\pi}^k X_n(s-),x)_H P_n(ds,dx) + 2\int_o^t \bar{\pi}^k X_n(s-) dX_n^s(s)$$

$$+ 2\int_o^t \int_{|x|_H\leq1} (\bar{\pi}^k X_n(s-),x)_H q_n(ds,dx) + <\bar{\pi}^k X_n^c>_t +$$

$$+ \int_o^t \int_{H\setminus\{o\}} |\bar{\pi}^k x|_H^2 P_n(ds,dx) .$$

From here by means of the lemma 2 of the paper [29] and the assumption 2) we obtain that

$$\lim_{k\to\infty} \overline{\lim_{n\to\infty}} \sup_{\sigma\in T(\mathbb{F}_n)} E_n |\bar{\pi}^k X_n(\sigma)|_H^2 = 0$$

and for each $T > 0$

$$\sup_n E_n[\int_o^T |X_n(s-)|_V^2 dN_s] < \infty .$$

At last from the assumption 3) we have that for each $\epsilon > 0$ and $k \geq 1$

$$\lim_{n\to\infty} P_n\{| (X_n(T_n+\delta_n),c_k)_H - (X_n(T_n),e_k)_H| > \epsilon\} = 0.$$

To complete the proof of the theorem 4 now it is enough to apply the criterion from the paper [30] and the known criterion of weakly relative compactness of probability measures in H (see [31]).

B. Grigelionis, R. Mikulevičius
Institute of Mathematics and Cybernetics
Lithuanian Academy of Sciences
University of Vilnius

Vilnius, Lithuania
U.S.S.R.

References

1. Galavotti, G.; Jona-Lasinio, G.: Limit theorems for multidimensional markovian processes. - Comm. Math. Phys., 1975, vol. 41, p. 301-307

2. Viot, M.: Solutions faibles d'équations aux dérivées partielles stochastiques non lineaires. - These doct. sci. Univ. P. et M. Curie, Paris, 1976.

3. Martin-Löf, A.: Limit theorems for motion of a Poisson system of independent markovian particles with high density. - Z. Wahr.verw. Geb., 1976, B. 34, S. 205-223

4. Dawson D. A. Stochastic evolution equations and related measure processes. - J. Multivariate Anal., 1975, vol. 5, No. 1, p. 1-52

5. Dawson, D. A.: Critical measure diffusion processes. - Z. Wahr.verw. Geb., 1977, B. 40, S. 125-145

6. Dawson, D. A.: Stochastic measure processes. - Stochastic Nonlinear Systems, ed. L. Arnold, R. Lefever, Springer, 1981, p. 185-199

7. Dawson, D. A.; Hochberg, K. J.: Wandering random measures in the Fleming - Viot model. - Carleton Math. Lecture Notes No. 33, 1981

8. Dawson, D. A.: Critical dynamics and fluctuations for a mean field model of cooperative behaviour. - Carleton Math. Lecture Note No. 33, 1981

9. Holley, R.; Stroock, D. W.: Generalized Ornstein - Uhlenbeck processes and infinite particle branching brownian motions. - RIMS Kyoto Univ., 1978, vol. 14, p. 741-788

10. Holley, R.; Stroock, D. W.: Central limit phenomena of various interacting systems. - Ann. of Math., 1979, vol. 110, p. 333-393

11. Holley, R.; Stroock, D.: Generalized Ornstein - Uhlenbeck processes as limits of interacting systems. - Proc. LMS Durham Symp., p. 152-168; Lecture Notes in Math. 851, Springer, 1981

12. Vishik, M. I.: Komech, A. I.; Fursikov A. V.:
    Mathematical problems of hydromechanics.
    - Uspechi Math. Nauk., 1979, vol. 34, No. 5,
    p. 135-210

13. Arnold, L.; Theodosopulu M.: Deterministic limit
    of the stochastic model of chemical reaction with
    diffusion. - Adv. Appl. Probab., 1980, vol. 12,
    p. 367-379

14. Arnold, L.: Mathematical models of chemical
    reactions. - Stochastic systems, Dordrecht, 1981

15. Kurtz, T. G.: Approximation of population
    processes. - CBMS - NSF Regional Conference
    series in Applied Mathematics, SIAM, 1981.

16. Tanaka, H.; Hitsuda, M.: Central limit theorem
    for a simple diffusion model of interacting
    particles. - Hirosima Math. J., 1981, vol. 11,
    p. 415-423

17. Hitsuda M. Central limit theorem for a simple
    interacting diffusion model and S'-valued
    processes. - Lecture Notes in Math., 1021,
    Proc. IV USSR-Japan Symp. Probab. Th. and
    Math. Stat., p. 238-242, Springer, 1983.

18. Kotelenez, P.: Law of large numbers and central
    limit theorem for chemical reaction with
    diffusion. - Dr. dissertation Univ. Bremen, 1982

19. Bouc R.; Pardoux, E.: Asymptotic analysis of
    PDEs with wideband driving noise. - Publ. Math.
    Appl. Marseille-Toulon, 83-4, 1983

20. Fouque, J.-P.: La convergence en loi pour les
    processus a valeurs dans un espace nucleaire.
    - Preprint, 1983, Paris

21. Bourbaki, N.: Integration, ch. IX. - Hermann,
    Paris

22. Grigelionis, B.; Mikulevičius, R.: On stably weak
    convergence of semimartingales and point
    processes. - Theory of Probab. and Appl., 1983,
    vol. XXVIII, No. 2, p. 320-332

23. Jacod, J.: Calcul stochastique et problemes de
    martingales. - Lecture Notes in Math. 714,
    Springer Verlag, 1979

24. Metivier, M.; Pellaumail, J.: Stochastic integration. - Academic Press, New York - London, 1980

25 Metivier, M.: Semimartingales. - W. de Gruyter, Berlin - New York, 1982

26. Krylov, N.V.; Rozovskii, B.L.: On stochastic evolution equations. - Modern Problems in Math., vol. 14, VINITI, M., 1979, p. 71-146

27. Gyöngy, I.; Krylov, N.V.: On stochastic equations with respect to semimartingales II, Ito formula in Banach spaces. - Stochastics, 1982, vol. 6, No. 3 + 4, p. 153-173

28. Lenglart E. Relation de domination entre deux processus. - Ann. Inst. H. Poincaré, 1977, vol. 13, p. 171-179

29. Grigelionis, B.; Mikulevičius, R.: Stochastic evolution equations and densities of the conditional distributions. - Lecture notes in Control and Inform. Theory, 49, Springer, 1983

30. Aldous, D.: Stopping times and tightness. - Ann. Probab., 1978, vol. 6, No. 2, p. 335-340

31. Parthasarathy, K. R.: Probability measures on metric space. - Academic Press, N.Y. - L., 1967

Michel Metivier

TIGHTNESS OF SEQUENCES OF HILBERT VALUED
MARTINGALES

Abstract

This paper gives a sufficient condition for the tightness
of a sequence $(M^n)_{n \in \mathbb{N}}$ of Hilbert valued martingales. This
result applies directly to some situations of "accompanying
martingales" as considered for example by L. Arnold,
M. Theodosopulu and P. Kotelenez.

1.   INTRODUCTION

Let us recall that if M is a $\mathbb{H}$-valued martingale (on
a stochastic basis $(\Omega, (\mathcal{F}_t)_{t \in [0,T]}, P)$), $\mathbb{H}$ being a separable Hilbert-space, there is a unique process denoted by
$\ll M \gg$ (see [6] chapter 4) with the following properties :
$\ll M \gg$ is the $\mathbb{H} \hat{\otimes}_1 \mathbb{H}$-valued process (matrix-valued process)
such that $M \otimes M - \ll M \gg$ is a martingale, with $\ll M \gg$ predictable and with finite variation. Let us recall that if (h_i) is
an orthonormal basis of $\mathbb{H}$, the elements of $\mathbb{H} \hat{\otimes}_1 \mathbb{H}$ are of
the form $\hat{y} := \sum_{i,j \in \mathbb{N}} \lambda_{ij} h_i \otimes h_j$ with $\| \hat{y} \|_1 = \sum |\lambda_{ij}| < \infty$,
and $\mathbb{H} \hat{\otimes}_1 \mathbb{H}$ is a Banach space for the norm $\| \hat{y} \|_1$. This space
is included in the Hilbert-Schmidt tensor product

$$\mathbb{H} \hat{\otimes}_2 \mathbb{H} := \{ \hat{y} : \hat{y} = \sum_{i,j} \lambda_{i,j} h_i \otimes h_j, \ \sum_{ij} |\lambda_{ij}|^2 < \infty \} .$$

L. Arnold and P. Kotelenez (eds.), Stochastic Space-Time Models and Limit Theorems, 217–226.
© *1985 by D. Reidel Publishing Company.*

The space $\mathbb{H} \hat{\otimes}_2 \mathbb{H}$ is a Hilbert space with $h_i \otimes h_j$ as an orthonormal basis and norm $\|\hat{y}\|_2 = (\sum_{ij} |\lambda_{ij}|^2)^{\frac{1}{2}}$. The injection from $\mathbb{H} \hat{\otimes}_1 \mathbb{H}$ into $\mathbb{H} \hat{\otimes}_2 \mathbb{H}$ is continuous.

We denote by $\lbrace\!\!\lbrace M \rbrace\!\!\rbrace$ the real valued increasing process :

(1.1) $\lbrace\!\!\lbrace M \rbrace\!\!\rbrace_t := \text{trace} \ll M \gg_t := \sum_i \ll M \gg_t^{ii}$.

In most practical problems the process $\ll M \gg$ has an integral representation of the type

(1.2) $\ll M \gg_t = \int_o^t \Phi(s,\omega)\,ds$

where Φ is a $\mathbb{H} \hat{\otimes}_1 \mathbb{H}$-valued process.

The theorem we present here (theorem 2.2) gives a tightness condition for a sequence $(M^n)_{n \in \mathbb{N}}$ of martingales involving only the processes $\ll M^n \gg$ and which applies very easily when these processes have the form (1.2).

Each process (M^n) is assumed to be defined on a stochastic basis $(\Omega^n, (\mathscr{F}^n_t)_{t \in [0,T]}, P^n)$.

2.   A TIGHTNESS CONDITION

An immediate extension of a result due to Rebolledo (see [9] or [8] for a short exposition) is the following.

2.1   Theorem (Rebolledo)

Let $(M^n)_{n \in \mathbb{N}}$ be a sequence of $\mathbb{H}$-valued right continuous martingales. We assume the following two properties [A] and [T] for this sequence :

[A] For every ε , $\eta > 0$ there exists $\delta > 0$ such that for every sequence (τ^n) of stopping times $(\tau^n$ is a stopping time on Ω^n with respect to the filtration $(\mathcal{F}_t^n)_{t \in [0,T]})$

$$\lim_{n \to \infty} \sup P^n \{ \langle\!\langle M^n \rangle\!\rangle_{\tau^n + \delta} - \langle\!\langle M^n \rangle\!\rangle_{\tau} > \eta \} \leqslant \varepsilon$$

[T] For every $t \in [0,T]$ and every $\varepsilon > 0$ there exists a compact set C in $\mathbb{H}$ such that

$$\sup_n P^n \{ M_t^n \notin C \} \leqslant \varepsilon \; .$$

Then, denoting by $\widetilde{P}^n$ the law of X^n in the Skorokhod space $\widetilde{\Omega} := \mathbb{D}(0,T ; \mathbb{H})$ of the process M^n , the sequence $(\widetilde{P}^n)_{n \in \mathbb{N}}$ is tight.

What we give now is a sufficient condition for [T] to be true. As said previously, this condition involves only the processes $\langle\!\langle M^n \rangle\!\rangle$.

2.2   Theorem

If the sequence $(\langle\!\langle M^n \rangle\!\rangle^{\frac{1}{2}})_{n \in \mathbb{N}}$ of $\mathbb{H} \overset{\wedge}{\otimes}_2 \mathbb{H}$ -valued processes satisfies [T] the same condition holds for the process $(M^n)_{n \in \mathbb{N}}$.

Proof. By assumption, for each t in [0,T] there exists a compact subset $\hat{C}$ of $\mathbb{H} \overset{\wedge}{\otimes}_2 \mathbb{H}$ such that

(2.2.1) $\sup_n P^n \{ \langle\!\langle M^n \rangle\!\rangle_t \notin \hat{C} \} \leqslant \varepsilon \; .$

Being compact the subset $\hat{C}$ is bounded and such that for every $\rho > 0$ there exists a finite dimensional subspace $\hat{\mathbb{F}}$ of $\mathbb{H} \hat{\otimes}_2 \mathbb{H}$ such that $\sup_{y \in \hat{C}} \| y - \Pi_{\hat{\mathbb{F}}} (y) \|^2 \leq \rho$, $\Pi_{\hat{\mathbb{F}}}$ denoting the orthogonal projection on $\hat{\mathbb{F}}$.

Our first claim is that the finite dimensional subspace $\hat{\mathbb{F}}$ can be taken of the form $\mathbb{H}_\rho \otimes \mathbb{H}_\rho$ where $\mathbb{H}_\rho$ is a finite dimensional subspace of $\mathbb{H}$. In fact if $\sum_{i,j} \xi^n_{ij} h_i \otimes h_j$ $n=1,\ldots,K$ is an orthonormal basis of a subspace $\hat{\mathbb{F}}$ of $\mathbb{H} \hat{\otimes}_2 \mathbb{H}$ such that $\sup_{y \in \hat{C}} \| y - \Pi_{\hat{\mathbb{F}}} (y) \|^2 \leq \rho/2$ and if we consider the subspaces $\hat{\mathbb{F}}_J$ generated by the vectors $\{ \sum_{i,j \in J} \xi^n_{ij} h_i \otimes h_j$, $n=1,\ldots,K\}$ for each finite subset J of $\mathbb{N}$

$$\lim_{J \uparrow} \| y - \Pi_{\hat{\mathbb{F}}_J} (y) \| = \| y - \Pi_{\hat{\mathbb{F}}} (y) \| \quad .$$

The compactness of $\hat{C}$ and the continuity of $y \to \Pi_{\hat{\mathbb{F}}}(y)$ implies the existence of J such that

(2.2.2) $\quad \sup_{y \in \hat{C}} \| y - \Pi_{\hat{\mathbb{F}}_J} (y) \| \leq \rho$.

But $\hat{\mathbb{F}}_J = \mathbb{H}_J \otimes \mathbb{H}_J$ where $\mathbb{H}_J$ is the subspace of $\mathbb{H}$ generated by $\{h_i : i \in J\}$.

Our second step will be to prove that for every $\varepsilon > 0$ and $\beta > 0$ we can find $\alpha > 0$ and a finite dimensional subspace $\mathbb{H}_\rho$ such that

(2.2.3) $\quad P^n \{\| M^n_t \| > \alpha \} \leq \varepsilon$

and

(2.2.4) $P^n\{\| M_t^n - \Pi_{H_\rho} M_t^n \|^2 > \beta\} \leq \epsilon$

It is easily seen that these properties will imply the hypothesis [T] for the sequence $(M_t^n)_{n \in \mathbb{N}}$ of $\mathbb{H}$-valued random variables.

To derive (2.2.3) we use the following consequence of a lemma of Lenglart (see [5] or [8]) : for every $b > 0$, $a > 0$ and every stopping time τ^n : .

(2.2.5) $P^n \{\sup_{s \leq \tau^n} \| M_s^n \|^2 > b^2\} \leq \frac{1}{b^2} E\{\{M^n\}_{\tau^n} \wedge a\} + P\{\{M^n\}_{\tau^n} \geq a\}$

But, since $\{M^n\}_t = \text{trace} \ll M^n \gg_t = \| \ll M^n \gg_t^{\frac{1}{2}} \|_{H.S}$, if we take $b := \sup_{y \in \hat{c}} \| y \|^2_{H.S}$ we see that

$$P^n\{\| M_t^n \| > \alpha\} \leq b/a^2 + \epsilon/2 \ .$$

Therefore α can be chosen for (2.2.3) to hold. To prove (2.2.4) let us take $\rho \leq \epsilon/2 \ \beta^2$ and consider the finite ·dimensional subspace F_ρ as considered above. Let G_ρ be the orthogonal subspace of $\mathbb{H}_\rho$, call Π_1 the orthonormal projection on G_ρ and Π_2 the orthogonal projection on $\mathbb{H}_\rho$. To simplify we set $Q := \ll M \gg_t$. Let us assume we are able to prove

(2.2.6) $\{\Pi_1 M\}_t \leq \| Q^{\frac{1}{2}} - \Pi_{H_\rho \otimes H_\rho} Q^{\frac{1}{2}} \|^2_{H.S}$

Applying again and using the fact that

$$P\{\| Q^{\frac{1}{2}} - \Pi_{H_\rho \otimes H_\rho} Q^{\frac{1}{2}}\|^2_{H.S} > \rho\} < \varepsilon/2$$

we obtain

$$P\{|\Pi_1 M|_t > \beta\} \leqslant \rho/\beta^2 + \varepsilon/2 .$$

From the choice of ρ we have

$$P\{|\Pi_1 M|_t > \beta\} \leqslant \varepsilon .$$

This is (2.2.3). The theorem will follow from the proof of (2.2.5). This is done in the next lemma.

2.3   Lemma

Let $Q := \ll M \gg_t$. Let $H_1 \oplus H_2$ be an orthogonal de-composition of H and $\Pi_{H_2 \hat{\otimes} H_2} Q^{\frac{1}{2}}$ the orthogonal projection of $Q^{\frac{1}{2}}$ on $H_2 \hat{\otimes}_2 H_2$. Let Π_1 be the orthogonal projection on H_2 . Then

$$\{\Pi_1 M\}_t \leqslant \| Q^{\frac{1}{2}} - \Pi_{H_2 \hat{\otimes}_2 H_2} Q^{\frac{1}{2}}\|^2_{H.S}$$

Proof. Call $\mathcal{Q}$ the operator in $\mathcal{L}(H ; H)$ associated to Q .

(2.3.1) $(\mathrm{Proj}_{H_i \hat{\otimes}_2 H_j} Q^{\frac{1}{2}}) = \Pi_i \circ Q^{\frac{1}{2}} \circ \Pi_j$ $i,j = 1,2.$

Let us write $\tilde{Q}$ for the Hilbert-Schmidt operator associa-ted to $Q \in H \hat{\otimes}_2 H$.

Note first that

(2.3.2) $\{\Pi_2 M\}_t = \mathrm{trace} \, \Pi_2 \circ \tilde{Q} \circ \Pi_2 .$

Then

$$\| \Pi_2 \circ \tilde{Q}^{\frac{1}{2}} \circ \Pi_2\|^2_{H.S} \leqslant \| \Pi_2 \circ \tilde{Q}^{\frac{1}{2}}\|^2_{H.S} = \mathrm{trace}(\Pi_2 \circ \tilde{Q}^{\frac{1}{2}} \circ \tilde{Q}^{\frac{1}{2}} \circ \Pi_2)$$
$$= \mathrm{trace}(\Pi_2 \circ \tilde{Q} \circ \Pi_2) .$$

Therefore, from (2.3.2)

(2.3.3) $\| \Pi_2 \circ \widetilde{Q}^{\frac{1}{2}} \circ \Pi_2 \|_{H.S}^2 \leqslant <\Pi_2 M>_t$.

Moreover, as a consequence of $\| M_t \|^2 = \| \Pi_1 M_t \|^2 + \| \Pi_2 M_t \|^2$

(2.3.4) $\langle M \rangle_t = <\Pi_1 M>_t + <\Pi_2 M>_t$.

From

$$\langle M \rangle_t = \sum_{i,j=1}^{2} \| \Pi_i \circ \widetilde{Q}^{\frac{1}{2}} \circ \Pi_j \|_{H.S}^2$$

we obtain easily the inequality of the lemma. ∎

The following proposition gives a sufficient condition of wide use to check the condition of Theorem 2.2.

2.4 Proposition

Let us assume

(i) $\lim_{\rho \to \infty} \sup_n P^n \{ \langle M^n \rangle_t > \rho \} = 0$

(ii)   there exists a complete orthonormal system
$(h_k)_{k \in \mathbb{N}}$ in $\mathbb{H}$ such that

$$\lim_{n \to \infty} \sup_n P^n \{ \sum_{k=m}^{\infty} (h_k) ; \ll \widetilde{M}^n \gg_t h_k) > \eta \} = 0 \ .$$

Then the sequence $(\ll M^n \gg_t^{\frac{1}{2}})_{n \in \mathbb{N}}$ is tight in $\mathbb{H} \overset{\wedge}{\otimes}_2 \mathbb{H}$.

Proof. It is easy to show the existence, for each
$\varepsilon > 0$ and $\eta > 0$, of a finite dimensional subspace L_ε
of $\mathbb{H} \overset{\wedge}{\otimes}_2 \mathbb{H}$ and a bounded set B_2 in $\mathbb{H} \overset{\wedge}{\otimes}_2 \mathbb{H}$ such that :

$$\sup_n P^n \{ \ll M^n \gg_t^{\frac{1}{2}} \notin B_\varepsilon \} \leqslant \varepsilon$$

and

$$\sup_{n} P^n \{ \| \ll M^n \gg_t^{\frac{1}{2}} - \Pi(\ll M^n \gg_t^{\frac{1}{2}} \|_2 > \eta \} \leqslant \varepsilon$$

where Π denotes the projection operator on L_ε .

3. EXAMPLE OF APPLICATION

In [4] the following sequence of "accompanying martin-gales" $(M^n)_{n \in \mathbb{N}}$ is considered.

$\mathbb{H}$ is a Sobolev space $\mathbb{H}_{-q}$ (see [4] for definitions) which is the dual of $\mathbb{H}_q \hookrightarrow L^2[0,1]$. The martingales M^n are such that for every $u, v \in \mathbb{H}_q$:

$$(3.1) \qquad <u ; \ll M^n \gg_t v> = \int_o^t <u ; a^n(X_s^n) v> ds$$

where the processes X^n take their values in $L^2[0,1]$ and $g \sim a^n(g)$ is an operator on $L^2[0,1]$ such that for every $u, v \in \mathbb{H}_1$ and $g \in L_\infty[0,1]$:

$$(a^n(g)u;v) \leqslant K \ r_n \ \| g \|_\infty \ \| u \|_1 \ \| v \|_1$$

If it is assumed, for example as in [3] , that on $[0,T]$ the paths of X^n are bounded in $L_\infty[0,1]$, we obtain, using the fact that the imbedding $i : H_2 \sim H_1$ is Hilbert-Schmidt:

$$(3.2) \qquad \ll \frac{1}{\sqrt{\varepsilon_n}} M^n \gg_t = \int_o^t \frac{1}{\varepsilon_n} a^n(X_s^n) \ ds$$

with

$$(3.3) \qquad \text{trace } (\frac{1}{\varepsilon_n} a^n(X_s^n)) \leqslant K$$

and for any complete orthonormal basis in H_2

$$(3.4) \qquad \lim_{m \to \infty} \sum_{k=m}^{\infty} <h_k, \frac{1}{\varepsilon_n} \ll M^n \gg_t h_k> \; \leqslant \; \lim_{m \to \infty} K \; T \sum_{k=m}^{n} \| h_k \|_1^2 = 0 \; .$$

Inequalities (3.3) and (3.4) easily imply the conditions of theorem 2.2 and proposition 2.3.

For more details on this type of applications the reader is referred to the forthcoming paper [7].

Acknowledgements. The author wants to thank the Center for Stochastic Processes of the University of North Carolina at Chapel Hill for supporting him under grant NO F4 620-82-C-0009 of the Air Force Office of Scientific Research during the summer 1983, while the results presented here were produced.

Professor Michel METIVIER
Centre de Mathématiques Appliquées -
Ecole Polytechnique
PALAISEAU CEDEX 91128
FRANCE

REFERENCES

[1]   Aldous, D.: 1978, "Stopping times and tightness",
          Ann. of Prob. 6, 2, 335-340.

[2]   Arnold, L.: 1981, "Mathematical models of chemical reac-
          tions" in M. Hazewinkel and J. Willems (ed.),
          Stochastic systems, Dordrecht.

[3]   Arnold, L. and Theodosopulu, M.: 1980, "Deterministic
          limit of the stochastic model of chemical reactions
          with diffusion", Adv. Appl. Prob., 1 , 363-329.

[4] Kotelenez, P.: 1982, Law of large numbers and central
        limit theorem for chemical reactions with diffusions
        Universität Bremen.

[5] Lenglart, E.: 1977, "Relations de domination entre deux
        processus", Ann. Inst. H. Poincaré, B XIII, 171-179.

[6] Métivier, M.: 1982, Semimartingales, De Gruyter, New York.

[7] Métivier, M.: 1984, "Convergence faible et principe
        d'invariance pour des martingales à valeurs dans
        des espaces de Sobolev", to appear in Ann. Inst.
        H. Poincaré.

[8] Métivier, M. and Joffe, A.: 1983, Weak convergence of
        sequences of semimartingales with applications to
        multitype branching processes, preprint.

[9] Rebolledo, R.: 1979, "La méthode des martingales appli-
        quée à la convergence en loi des processus",
        Mémoires de la S.M.F., 62.

E. Pardoux

ASYMPTOTIC ANALYSIS OF A SEMI-LINEAR
PDE WITH WIDE-BAND NOISE DISTURBANCES

ABSTRACT. We present an asymptotic analysis in the "white-noise limit" of a semi-linear parabolic partial differential equation, whose coefficients are perturbed by a wide-band noise. We prove the convergence in law towards the solution of an Ito stochastic PDE, thus generalizing the results in [3] for linear PDEs.

1. INTRODUCTION

We study here the "white noise limit" of a PDE, whose coefficients are perturbed by a wide band noise. We will obtain an Ito-type stochastic non linear PDE for the limit in law, as $\epsilon \to o$, of u^{ϵ}, the solution of :

$$\frac{\partial u^{\epsilon}}{\partial t}(t,x) + Au^{\epsilon}(t,x) + F(u^{\epsilon}(t,x)) + \frac{1}{\epsilon} B(Z_{t/\epsilon^2}) u^{\epsilon}(t,x) =$$

$$= f(x) + \frac{1}{\epsilon} g(Z_{t/\epsilon^2}, x)$$

$t \geqslant o,\ x \in D \subset \mathbb{R}^n$; $u^{\epsilon}(t,x) = o,\ x \in \partial D;\ u^{\epsilon}(o,x) = u_o(x)$

where $\forall z \in \mathbb{R}^d$, A and B(z) are linear partial differential operators, respectively of second and first order, $\{Z_t\}$ is a stationary positive recurrent diffusion process with values in $\mathbb{R}^d$ which perturbs the coefficients of B and g, with $\quad E[B(Z_t)u(x)] = o,\quad E[g(Z_t,x)] = o$

for all u in some function space, and F is a Lipschitz

227

L. Arnold and P. Kotelenez (eds.), Stochastic Space-Time Models and Limit Theorems, 227–242.
© 1985 by D. Reidel Publishing Company.

mapping from $\mathbb{R}$ into $\mathbb{R}$.

Similar convergence results in the finite dimensional case – i.e. where (1.1) is replaced by an ODE – can be formd e.g. in PAPANICOLAOU [8], BLANKENSHIP-PAPANICOLAOU [2], KUSHNER [6]. This paper generalises the results of BOUC-PARDOUX [3], where $F \equiv o$ (i.e. the PDE is linear). Convergence results for linear PDEs have also been obtained recently by KUSHNER and HUANG [7],[4].

Since the arguments are very similar to those in [3], we will only sketch here most of the proofs, giving some details concerning the treatment of the non-linear term F.

In section 2, we state our hypotheses concerning the diffusion process $\{Z_t\}$, and recall the results in [3] about its associated Poisson equation, which will be slightly generalized. In section 3, we formulate the precise assumptions on A, B, F, f, g, and prove the convergence of the law Q^ε of u^ε, towards the solution of a martingale problem.

2. THE RECURRENT DIFFUSION $\{Z_t\}$ AND ITS ASSOCIATED POISSON EQUATION.

2.1 The diffusion process $\{Z_t\}$.

$\{Z_t\}$ is supposed to be a diffusion process with values in $\mathbb{R}^d$, defined on a probability space (Ω_0, F_0, P), whose infinitesimal generator is given by :

$$ L = \frac{1}{2} \bar{a}_{ij}(z) \frac{\partial^2}{\partial z_i \partial z_j} + \bar{b}_i(z) \frac{\partial}{\partial z_i} $$

Note that we use here and henceforth the convention of summation upon repeated indices. We now make some hypotheses on the coefficients $\bar{a}$ and $\bar{b}$:

(H1) $\bar{a}_{ij}$, $\bar{b}_i$, $a_i \triangleq \bar{b}_i - \frac{1}{2} \frac{\partial \bar{a}_{ij}}{\partial z_j}$ are bounded C^∞ functions from $\mathbb{R}^d$ into $\mathbb{R}$; a_{ij} and b_i are uniformly Hölder continuous with some exponent $\alpha \in]o,1]$; $i,j = 1,\ldots,d$.

(H2) $\exists \beta > o$ s.t. $\bar{a}_{ij}(z) u_i u_j \geq \beta |u|^2$, $\forall z$, $u \in \mathbb{R}^d$

(H3) $\exists N > 0$ s.t. $\bar{b}_i(z)z_i|z|^{-1} \leqslant -N^{-1}$, $\forall z$ s.t. $|z| \geqslant N$

      It follows from (H3)- see KHASMINSKII [5] - that the diffusion process $\{Z_t\}$ is positive recurrent. It then follows from (H1)-(H2) that its unique invariant probability measure possess a smooth density $p(x)$, s.t. $p(x) > 0$, $\forall x \in \mathbb{R}^d$. From now on, we suppose that this invariant measure is the law of Z_0.
      Assume moreover :

(H4) $\exists N > 0$ s.t. $[|z|p(z)]^{-1}z_i \dfrac{\partial}{\partial z_j} [\bar{a}_{ij} p](z) \leqslant -N^{-1}$, $\forall z$ s.t.

$$|z| \geqslant N$$

Remark 2.1 (H4) is equivelent to the more explicit condition (H3) whenever :

$$z_i[(2p(z))^{-1} \dfrac{\partial}{\partial z_j}(\bar{a}_{ij} p)(z) - \bar{b}_i(z)] \equiv 0$$

The latter equality holds in particular when the process $\{Z_t\}$ is "time-reversible", i.e. when
$\forall t > 0$, $\forall \phi \in C_b(\mathbb{R}^d), E[\phi(Z_t)/Z_0 = x] = E[\phi(Z_0)/Z_t = x]$, which is always the case if $d = 1$. $\square$

Remark 2.2 We do not assume here, as we did in [3], that $p^{-1}a_{ij} \dfrac{\partial p}{\partial z_j}$ be bounded over $\mathbb{R}^d$, $\forall i = 1,\ldots,d$. This does not affect the proof of Theorem 2.3 below, but we will have to adapt the proof of Theorem 2.6. $\square$

2.2 Solution of the associated Poisson equation.

Define $\hat{L}^2 \overset{\Delta}{=} L^2(\mathbb{R}^d; p(z)dz)$, $\hat{L}_0^2 = \{u \in \hat{L}^2; \int p(z)u(z)dz = 0\}$, and

$\hat{H}^1 \overset{\Delta}{=} \{u \in \hat{L}^2; \dfrac{\partial u}{\partial z_i} \in \hat{L}^2, i=1,\ldots,d\}$. $(.,.)_\wedge$ and $|.|_\wedge$ will denote respectively the scalar product and norm of $\hat{L}^2$.
      We consider L as an unbounded operator on $\hat{L}^2$, and denote $D(L) \overset{\Delta}{=} \{u \in \hat{L}^2; Lu \in \hat{L}^2\}$.
Note that for $u \in D(L)$,

$$-(Lu, u)_\wedge = (\mathcal{L}u, u)_\wedge$$

where $\mathcal{L}$ is a self-adjoint non-negative operator on $\hat{L}^2$

defined by :
$$\mathcal{L} = - \frac{1}{2p} \frac{\partial}{\partial z_j} [\bar{a}_{ij} \, p \, \frac{\partial}{\partial z_i}(.)]$$

In fact, $\mathcal{L} = L$ if and only if $\{Z_t\}$ is time-reversible. From (H2), one can deduce :

$$(\mathcal{L} u, u)_\wedge \geqslant \frac{\beta}{2} \sum_{i=1}^{d} |\frac{\partial u}{\partial z_i}|_\wedge^2, \ \forall u \in D(L)$$

It then follows that the eigenspace associated with the eigenvalue O of $\mathcal{L}$ is the set of constants. We now quote several results from [3]:

Theorem 2.3 Assume that (H1),(H2) and (H4) hold. Then o is an isolated point of the spectrum of $\mathcal{L}$. $\Box$
We note that, as mentionned in [3], the boundedness of the drift b is not necessary for Theorem 2.3 to hold.

Corollary 2.4 Under the assumptions of Theorem 2.3, if $f \in \hat{L}_o^2$, then there exists a unique solution $v \in \hat{H}^1 \cap \hat{L}_o^2$ to the Poisson equation :
$$L v + f = o$$
Moreover, $v(z) = \int_o^\infty E_z[f(Z_t)]dt$, where E_z denotes expectation under the law of $\{Z_t\}$ conditionned upon $Z_o = z$. $\Box$

Remark 2.5 It follows from Theorem 2.3 that there exists $\lambda > o$ s.t. $\forall f \in \hat{L}_o^2$ (i.e. orthogonal in $\hat{L}^2$ to the eigenspace associated to o), $|E.f(Z_t)|_\wedge \leqslant e^{-\lambda t}|f|_\wedge$, which insures the convergence of the integral $\int_o^\infty E_z[f(Z_t)]dt$. Note that the above inequality, with the $\hat{L}^2$-norm replaced by the sup-norm, would follow from Doeblin's condition, which is usually assumed for recurrent processes with values in a compact set, but can't be satisfied by a diffusion process with values in the whole space $\mathbb{R}^d$. $\Box$
We will need conditions on f which insure that v be bounded.

Theorem 2.6 Assume (H1), (H2), (H3) and (H4).
Let $f \in \hat{L}_o^2 \cap L^\infty(\mathbb{R}^d)$ satisfying the following condition :

$$\exists\, N > 0 \quad s.t. \quad \int_{N}^{+\infty} \text{ess} \sup_{|z| \geq t} |f(z)|\, dt < \infty$$

Then the solution $v \in \hat{L}_0^2 \cap \hat{H}_0^1$ of $Lv + f = 0$ is a.e. bounded and $\exists\, c\,(N)$ s.t.

$$\|v\|_\infty \leq c(N)\,[\|f\|_\infty + \int_{N}^{\infty} \text{ess} \sup_{|z| \geq t} |f(z)|\, dt] \quad (*)$$

where $\|.\|_\infty$ denotes the sup norm.

<u>Proof</u> : This is Theorem 1.12 of [3], which is due to P.L. <u>LIONS</u>. The proof given there uses the boundedness of

$p^{-1}\bar{a}_{ij}\dfrac{\partial\, p}{\partial z_j}$. Now, from the smoothness and strict positivity

of p, this quantity is still locally bounded. Therefore, instead of integrating by parts the quantity

$$(L\bar{w}, \bar{w}^-)_\Lambda$$

we integrate by parts :

$$(L\bar{w}, \varphi\bar{w}^-)_\Lambda$$

where φ is C^∞ with compact support, and ≥ 0.
We obtain :

$$0 \leq a_{B_0} (\bar{w}^-, \varphi\bar{w}^-) = -\tfrac{1}{2}(\bar{a}_{ij}\,\dfrac{\partial \bar{w}^-}{\partial z_i}, \varphi\dfrac{\partial \bar{w}^-}{\partial z_j})_\Lambda -$$

$$- (\bar{a}_{ij}\,\dfrac{\partial \bar{w}^-}{\partial z_i}, \dfrac{\partial \varphi}{\partial z_j}\bar{w}^-)_\Lambda - \tfrac{1}{4}(\bar{a}_{ij}\,\dfrac{\partial^2 \varphi}{\partial z_i \partial z_j}\bar{w}^-, \bar{w}^-)_\Lambda -$$

$$- \tfrac{1}{2}(\bar{a}_i\,\bar{w}^-, \dfrac{\partial \varphi}{\partial z_i}\bar{w}^-)_\Lambda$$

It is then easy to take the limit as $\varphi \to 1$, giving the same conclusion as in [3]. Finally, the estimate (*) follows from a careful analysis of the proof. $\square$

For $f \in \hat{L}_0^2$, we denote by Mf the element of $\hat{L}_0^2 \cap \hat{H}_0^1$ defined by:

$$(M f)\,(z) = \int_{0}^{\infty} E_z\, f\,(Z_t)\, dt$$

i.e. $M = - L^{-1}$.

Lemma 2.7 Let $f \in \hat{L}_o^2(\mathbb{R}^d) \cap C_b^1(\mathbb{R}^d)$ [i.e. f is of C^1-class, and is bounded together with its first partial derivatives]. Suppose that $Mf \in L^\infty(\mathbb{R}^d)$.
Then $\frac{\partial}{\partial z_i} Mf \in L^\infty(\mathbb{R}^d)$, $i=1,\ldots,d$; and $\exists c$ s.t.:

$$\| \frac{\partial Mf}{\partial z_i} \|_\infty \leqslant c \, [\, \| \frac{\partial f}{\partial z_i} \|_\infty + \| Mf \|_\infty] \quad \square$$

Corollary 2.8 There exists a constant $c(N)$, which depends only on N and the operator L, s.t.

$$\| \frac{\partial Mf}{\partial z_i} \|_\infty \leqslant c(N) \, [\| \frac{\partial f}{\partial z_i} \|_\infty + \| f \|_\infty + \int_N^\infty \operatorname{ess \, sup}_{|z| \geqslant t} |f(z)| \, dt]$$
$$\square$$

Let us make some comments in order to motivate what follows. Let first $f \in L^1(\mathbb{R}^d; p(z)dz)$. We have the following law of large numbers, which follows from a classical ergodic theorem :

$$\frac{1}{t} \int_0^t f(Z_s) ds \longrightarrow \int f(z) \, p(z) dz \quad \text{a.e.}$$

One might expect a central limit theorem of the following type to hold :

If $f \in \hat{L}_o^2$, $\frac{1}{\sqrt{t}} \int_0^t f(Z_s) ds \xrightarrow{\mathcal{L}} N(o, 2(f, Mf)_\wedge)$

where $Mf(z) = \int_0^\infty E_z f(Z_t) dt$. In fact, one might expect - see

BHATTACHARYA [1], TOUATI [10] – a functional central limit theorem of the type :

If $f \in \hat{L}_o^2$, $\frac{1}{\varepsilon} \int_0^{\varepsilon^2 t} f(Z_s) ds \xrightarrow[\varepsilon \to o]{\mathcal{L}}$ Wiener process with

covariance coefficient $2(f, Mf)_\wedge$. In fact, a very particular case of the proof in section 3 establishes the above functional central limit theorem for f satisfying the conditions of Theorem 2.6.

Note that :

$$\frac{1}{\varepsilon} \int_0^{\varepsilon^2 t} f(Z_s) ds = \frac{1}{\varepsilon} \int_0^t f(Z_{s/\varepsilon^2}) ds$$

therefore, one might say, in a sense, that $\{\frac{1}{\varepsilon} f(Z_{t/\varepsilon^2})\}$

converges to a white noise, the time – derivative of a Wiener process. This formal statement may give some intuition for the convergence result of section 3.

3. CONVERGENCE OF THE PARABOLIC PDE WITH WIDE BAND NOISE DISTURBANCES.

3.1 Hypotheses and Notations

D will denote an open subset of $\mathbb{R}^n$, and $H_o^1(D)$ the closure of the set of C^∞ functions with compact support in D, with respect to the norm :

$$\| u \| = \left(|u|^2 + \sum_{i=1}^{n} \left| \frac{\partial u}{\partial x_i} \right|^2 \right)^{1/2}$$

where $|.|$ denotes the usual norm in $L^2(D)$. Note that if the boundary ∂D of D is somewhat regular, $H_o^1(D)$ is exactly the subset of $H^1(D) \triangleq \{u \in L^2(D); \frac{\partial u}{\partial x_i} \in L^2(D); \ i=1,..,n\}$ consisting of those u which trace on ∂D is zero. We now identify $L^2(D)$ with its dual space, and the dual space of $H_o^1(D)$ with $H^{-1}(D)$. We denote by $< .,. >$ the pairing between $H_o^1(D)$ and $H^{-1}(D)$.

Suppose ave are given :

(G.1)
$$\begin{cases} A \in \mathcal{L}(H_o^1(D), H^{-1}(D)) \ s.t. \ \exists \, \alpha > o \ , \ \lambda \ \text{with} \ : \\[2mm] < A \, u, u > + \, \lambda |u|^2 \geq \alpha \, \|u\|^2 \ , \quad \forall u \in H_o^1(D) \end{cases}$$

A could be $- \Delta = - \sum_{i=1}^{n} \frac{\partial^2}{\partial x_i^2}$, or more generally an elliptic second order operator.

For any $z \in \mathbb{R}^d$, B(z) is a first order operator :

$$B(z) = b_i(x,z) \frac{\partial}{\partial z_i} + b_o(x,z)$$

(G.2) b_i is C^∞ in (x,z), the function and all its derivatives being bounded; $i=o,1,..,n$.

(G.3)
$$\begin{cases} \exists \, N > o \ s.t. \ \int_N^\infty \text{ess sup}(\sum_{|z| \geq t} \ |j| \leq d_i \ \left| \frac{\partial^{|j|} b_i(x,z)}{\partial x_1^{j_1}...\partial x_n^{j_n}} \right|) dt \leq N \ ; \\[4mm] i = o,1,..,n; \ \text{with} \ d_o=2, \ d_i=3, \ i=1,..,n. \end{cases}$$

(G.4) $\int b_0(x,z)p(z)dz = \int b_i(x,z)p(z)dz = o \quad \forall x \in D, \ i=1,..,d$

(G.5) $\begin{cases} F: D \times \mathbb{R} \to \mathbb{R} \text{ is a measurable mapping satisfying :} \\ \quad (i) \quad F(x,o) = o \\ \quad (ii) \quad |F(x,u)-F(x,v)| \leqslant k \ |u-v| \end{cases}$

In the sequel, for $u \in L^2(D)$, we will denote by $F(u)$ the element of $L^2(D)$ defined by $F(u)(x)=F(x,u(x))$.

(G.6) $\quad f \in L^2(D)$

(G.7) $\begin{cases} g: D \times \mathbb{R}^d \to \mathbb{R} \text{ is of class } C^1, \text{ and the following} \\ \text{functions of } x \text{ are supposed to belong to } L^2(D): \\ \sup_z |g_i(.,z)|, \sup_z |\dfrac{\partial g_i}{\partial z_j}(.,z)|, \int_N^\infty \text{ess } \sup_{|z| \geqslant t} |g_i(.,z)|dt \\ \text{for } i=o,1,..,n; \ j=1...d; \text{ where } g_o=g; g_i=\dfrac{\partial g}{\partial x_i}, i=1..n \end{cases}$

(G.8) $\int g(x,z)p(z)dz = o, \quad \forall x \in D$

(G.9) $\quad u_o \in L^2(D)$

For each $\varepsilon > o$, we define $u_t^\varepsilon(\omega_o,x)$ as the unique solution of :

$$\left. \begin{array}{c} \dfrac{du_t^\varepsilon}{dt} + A u_t^\varepsilon + F(u_t^\varepsilon) + \dfrac{1}{\varepsilon}B(Z_t^\varepsilon)u_t^\varepsilon - f + \dfrac{1}{\varepsilon}g(Z_t^\varepsilon) \\ u_o^\varepsilon = u_o \end{array} \right\} \quad (3.1)$$

which belongs P a.s. to $L^2(o,T;H_o^1(D)) \cap C([o,T];L^2(D)), \forall T > o.$ Here and in the sequel, $Z_t^\varepsilon \triangleq Z_{t/\varepsilon^2}$. Clearly, the operator

and forcing terms in (3.1) could depend on t; we have avoided that in order to simplify the notations. In the next sections all hypotheses (H.1)-(H.4) and (G.1)-(G.9) are assumed to hold.

3.2 Weak compactness

From now on, we fix an arbitrary $T > o$. $\overline{L^2(D)}$ will denote the space $L^2(D)$ endowed with its weak topology. We define :

$$\Omega = C([o,T]; \overline{L^2(D)}) \cap L^2(o,T; H_o^1(D))$$

and endow Ω with the supremum of the topology of uniform convergence on $C\{[o,T]; \overline{L^2(D)})$, and the weak topology on $L^2(o,T; H_o^1(D))$. Let F denote the Borel σ-field on Ω, and Q^ε the law $\{u_t^\varepsilon, t \in [o,T]\}$ on (Ω,F). We point out that also Ω is not a Polish space, Prohorov criterion for weak relative compactness of sets of probability measures on Ω holds-see VIOT [11].
On can prove-exactly as in [3]:

<u>Theorem 3.1</u>   The family of probability measures $\{Q^\varepsilon, \varepsilon > o\}$ on (Ω,F) is tight.

<u>Sketch of the proof</u> : We first note that a sufficient condition for $K \subset \Omega$ to be precompact is :

$$\sup_{u \in K} \sup_{t \in [o,T]} |u(t)| < \infty \qquad (i)$$

For each θ in a dense subset of $L^2(D)$, $\{t \rightarrow (u(t),\theta), u \in K\}$ is a set of equicontinuous mappings. $\qquad (ii)$

$$\sup_{u \in K} \int_o^T \|u(t)\|^2 dt < \infty \qquad (iii)$$

This indicates us that in order to prove the theorem, we need to show that $\exists c$ s.t:

$$E(\sup_{t \leq T}|u_t^\varepsilon|^2 + \int_o^T \|u_t^\varepsilon\|^2 dt) \leq c, \forall \varepsilon > o \qquad (3.2)$$

and also to take care of the modulus of continuity of (u_t^ε,θ), for θ in a dense subset of $L^2(D)$. The difficulty in establishing the estimate (3.2) comes from the $\frac{1}{\varepsilon}$ terms in the equation (3.1). The trick to kill them is to replace the quantity $|u_t^\varepsilon|^2$ to be estimated by :

$$|u_t^\varepsilon|^2 - 2\varepsilon(MB(Z_t^\varepsilon)u_t^\varepsilon,u_t^\varepsilon) + 2\varepsilon(Mg(Z_t^\varepsilon),u_t^\varepsilon)$$

where $M = - L^{-1}$ has been defined in §2. Applying Ito's formula to the above quantity, we obtain :

$$|u_t^\varepsilon|^2 - 2\varepsilon(MB(Z_t^\varepsilon)u_t^\varepsilon,u_t^\varepsilon) + 2\varepsilon(Mg(Z_t^\varepsilon),u_t^\varepsilon) +$$

$$+ 2\int_o^t <\tilde{A}(u_s^\varepsilon),u_s^\varepsilon> ds - 4\varepsilon\int_o^t <\tilde{A}(u_s^\varepsilon),M\tilde{B}(Z_s^\varepsilon)u_s^\varepsilon> ds +$$

$$+ 2\varepsilon \int_0^t <\tilde{A}(u_s^\varepsilon), Mg(Z_s^\varepsilon)> ds - 4 \int_0^t (M\tilde{B}(Z_s^\varepsilon)u_s^\varepsilon, B(Z_s^\varepsilon)u_s^\varepsilon) ds +$$

$$+ 2\int_0^t (B(Z_s^\varepsilon)u_s^\varepsilon, Mg(Z_s^\varepsilon)) ds = |u_0|^2 - 2\varepsilon(M\tilde{B}(Z_0)u_0, u_0) + 2\varepsilon(Mg(Z_0)u_0 +$$

$$+ 2\int_0^t (f, u_s^\varepsilon) ds + 2\varepsilon \int_0^t (Mg(Z_s^\varepsilon) - 2M\tilde{B}(Z_s^\varepsilon)u_s^\varepsilon, f) ds -$$

$$- 4\int_0^t (M\tilde{B}(Z_s^\varepsilon)u_s^\varepsilon, g(Z_s^\varepsilon)) ds + 2\int_0^t (Mg(Z_s^\varepsilon), g(Z_s^\varepsilon)) ds +$$

$$+ 2\int_0^t [(\nabla_z Mg(Z_s^\varepsilon), u_s^\varepsilon) - (\nabla_z MB(Z_s^\varepsilon)u_s^\varepsilon, u_s^\varepsilon)] \sigma(Z_s^\varepsilon) dW_s^\varepsilon$$

where $\tilde{A}(u) = Au + F(u)$, $\tilde{B} = \frac{1}{2}(B+B^*)$ is a zero order operator.
(3.2) now follows by standard PDE arguments, using the hypo-
theses, in particular (G.1). The rest of the proof is
identical to that in [3]. □

3.2   The limiting law

We now define the limiting law Q on (Ω, F).
Define $\bar{A} = A - \int_0^\infty E[B(Z_t)B(Z_0)] dt$

$$\bar{f} = f - \int_0^\infty E[B(Z_t)g(Z_0)] dt$$

Note that the differences between A and $\bar{A}$, f and $\bar{f}$ can be
interpreted as Ito-Stratonovich correcting terms.
For $u \in H_0^1(D)$, let $R(u) \in \mathcal{L}(L^2(D))$ be given by :

$$(R(u)h, k) = \int_0^\infty E[(B(Z_0)u - g(Z_0), h)(B(Z_t)u - g(Z_t), k) +$$

$$+ (B(Z_t)u - g(Z_t), h)(B(Z_0)u - g(Z_0), k)] dt, \quad h, k \in L^2(D).$$

For $\omega \in \Omega$, we denote $v_t(\omega) = \omega(t)$. We will say that a probability measure Q on (Ω, F) solves the martingale problem
$MP(u_0, \bar{A}, F, \bar{f}, R)$ if :

   (i) $v_0(\omega) = u_0$, Q a.s.

   (ii) $\forall \theta \in H_0^1(D)$, the process $\{M_t^\theta; t \in [o,T]\}$ given by :

$$M_t^\theta \triangleq (v_t, \theta) - (u_0, \theta) + \int_0^t <\bar{A} v_s + F(v_s) - \bar{f}, \theta> ds$$

is a continuous Q-martingale whose quadratic variation
process is given by :

$$<M^\theta>_t = \int_o^t R(v_s)ds$$

By the same argument as in [3], one can show that
$\exists K(u) \in \mathcal{L}(L^2(D))$ such that $u \to K(u)$ is linear, and
$R(u) = K(u) \circ K(u)^*$.

It then follows that, if Q solves $MP(u_o, \overline{A}, F, \overline{f}, R)$, there
exists a Q-cylindrical brownian motion
$\{W_t, t \in [o,T]\}$ on $L^2(D)$ [$\forall h \in L^2(D), (W_t, h)$ is a Wiener
process with associated increasing process $t|h|^2$], such that:

$$v_t + \int_o^t [\overline{A}v_s + F(v_s)]ds = u_o + \int_o^t \overline{f}\,ds + \int_o^t K(v_s)dW_s \qquad (3.3)$$

From [9], (3.3) has a unique strong solution.
It then follows – see VIOT [I'l] – that the martingale problem
$MP(u_o, \overline{A}, F, \overline{f}, R)$ is well-posed. In the sequel, Q will denote
the unique solution of $MP(u_o, \overline{A}, F, \overline{f}, R)$.

Remark 3.2   In some cases, the martingale in the right-hand
side of (3.3) can be written as an integral with respect to
a finite-dimensional Wiener process. Let us consider the
simplest example. Suppose that $B(z) = \alpha(z)B$ and $g(z) = \alpha(z)g$.

Then $(R(u)h, k) = 2(\int_o^\infty E[\alpha(Z_o)\alpha(Z_t)]dt)(Bu - g, h)(Bu - g, k)$.

Suppose (this is not a restriction) that $2\int_o^\infty E[\alpha(Z_o)\alpha(Z_t)]dt = 1$.
Then (3.3) reads :

$$v_t + \int_o^t (A - \frac{1}{2}B^2)v_s\,ds + \int_o^t F(v_s)ds = u_o + \int_o^t (f - \frac{1}{2}Bg)ds + \int_o^t (Bv_s - g)dW_s$$

where here $\{W_t\}$ is a one dimensional standard Wiener
process. $\square$

3.3. Convergence of Q^ε

It remains to show that Q is the limit of any converging
subsequence $\{Q^{\varepsilon_n}, n \in \mathbb{N}\}$, such that $\varepsilon_n \to o$. Here comes the
difficulty with the non-linear term F. Indeed, the mapping :

$$v_\cdot \quad \to \quad F(v_\cdot)$$

is not continuous on Ω, since a continuous non linear mapping from $L^2(D)$ into itself is usually not weakly continuous. Our proof will follow the lines of that of VIOT [11, Thm II 1.1].

Define $\Omega' = L^2(]o,T[\times D)$, which we endow with its weak topology, $\overline{\Omega} = \Omega \times \Omega'$ and $\overline{F}$ the Borel σ-field on $\overline{\Omega}$. Let $\overline{Q}^\varepsilon$ denote the law of $\{u_t^\varepsilon, F(u_t^\varepsilon); t \in [o,T]\}$ on $(\overline{\Omega},\overline{F})$. The tightness of $\{\overline{Q}^\varepsilon, \varepsilon > o\}$ follows immediately from that of $\{Q^\varepsilon, \varepsilon > o\}$. Define $r_t(\omega') = \omega'(t)$; By the argument of Theorem 2.8 in [3], we get that if $\overline{Q}$ is a weak-limit of a sequence $\{\overline{Q}^{\varepsilon_n}, n \ \text{IN}\}$, with $\varepsilon_n \to o$, then :

(i) $v_0(\omega) = u_0 \quad \overline{Q}$ a.s.

(ii) $\forall \theta \in H_o^1(D)$, the process $\{\overline{M}_t^\theta(\omega,\omega'), t \in [o,T]\}$

given by :

$$\overline{M}_t^\theta \overset{\Delta}{=} (v_t,\theta)-(u_o,\theta)-\int_o^t < \overline{A}\,v_s + r_s - \overline{f},\theta > ds$$

is a continuous $\overline{Q}$-martingale whose quadratic variation process is given by :

$$< \overline{M}^\theta >_t = \int_o^t R(v_s)ds$$

Clearly, since we are working on Lusin spaces, we have the desintegration :

$$Q(d\omega,d\omega') = S(d\omega)\,\mu_{(\iota)}(d\omega')$$

Suppose for a moment that we have established :

Lemma 3.3 $\forall w \in L^2(o,T;H_o^1(D))$, S a.s.

$$\int_{\Omega'} (\int_o^T (r_t(\omega'),w(t))dt)\mu_\omega(d\omega') = \int_o^T (F(v_t(\omega)),w(t))\,dt \quad \square$$

It then follows that the process:

$$M_t^\theta(\omega) \overset{\Delta}{=} \int \overline{M}_t^\theta(\omega,\omega')\mu_\omega(d\omega')$$

$$= (v_t(\omega),\theta)-(u_o,\theta)-\int_o^t < \overline{A}v_s(\omega)+F(v_s(\omega))- \overline{f},\theta > ds$$

is a continuous S-martingale, whose quadratic variation process is the same as that of $\overline{M}_t^\theta$. Then $S = Q$, the unique solution of $MP(u_o,\overline{A},F,\overline{f},R)$.
It is now easy to conclude :

<u>Theorem 3.4</u> Let $\varepsilon_n \to o$ be such that the sequence $\{Q^{\varepsilon_n}, n \in \mathbb{N}\}$ converges weakly. Then $Q^{\varepsilon_n} \Rightarrow Q$

<u>Proof</u> : We can extract a further subsequence $\{\varepsilon_m\}$, such that $\{\overline{Q}^{\varepsilon_m}, m \in \mathbb{N}\}$ converges weakly, as probability measures on $(\overline{\Omega}, \overline{F})$. From the above arguments,

$$\overline{Q}^{\varepsilon_m} \Rightarrow \overline{Q}$$

Then $\quad Q^{\varepsilon_m} = \int_{\Omega'} \overline{Q}^{\varepsilon_m}(., d\omega') \Rightarrow \int_{\Omega'} \overline{Q}(., d\omega') = Q$

<u>Proof of Lemma 3.3</u> $\overline{E}, \overline{E}^\varepsilon$ will denote respectively expectation with respect to $\overline{Q}, \overline{Q}^\varepsilon$. Let $\lambda > o$. For $w \in \mathscr{V} = L^2(o, T; H^1_o(D))$, define :

$$X(\omega, \omega', w) \triangleq \int_o^T e^{-\lambda t} \{ 2 < \overline{A}(v_t - w_t) + r_t - F(w_t), v_t - w_t > +$$

$$+ \lambda |v_t - w_t|^2 - \mathrm{TrR}(v_t - w_t) \} dt$$

Since $r_t(\omega') = F(v_t(\omega))$ $\quad \overline{Q}^\varepsilon$ a.s., it follows from our standing hypotheses that we can choose λ large enough such that :

$$\overline{E}^\varepsilon(X) \geqslant o, \forall w \in \mathscr{V} \qquad (3.4)$$

From now on, we fix λ such that (3.4) holds.

$$X = X_1 + X_2 \qquad , \text{ where}$$

$$X_1 = \int_o^T e^{-\lambda t} \{ 2 < \overline{A}v_t + r_t, v_t > + \lambda |v_t|^2 - \mathrm{TrR}(v_t) \} dt$$

Since $X_2 = X - X_1$ is continuous on $\overline{\Omega}, \forall w \in \mathscr{V}$, and $\overline{E}^\varepsilon(X_2^2) \leqslant c \ \forall \varepsilon > o$, it is easy to show that, provided $\overline{Q}^{\varepsilon_n} \Rightarrow \overline{Q}$,

$$\overline{E}^{\varepsilon_n}(X_2) \to \overline{E}(X_2) \qquad (3.5)$$

We now look for another expression for $\overline{E}^\varepsilon(X_1)$. A slight modification of the computation made in the proof of Theorem 3.1 leads to :

$$e^{-\lambda T}|u_T^\varepsilon|^2 + 2 \int_o^T e^{-\lambda t} < \tilde{A}(u_t^\varepsilon) - MB(Z_t^\varepsilon)B(Z_t^\varepsilon)u_t^\varepsilon, u_t^\varepsilon > dt +$$

$$+ \lambda \int_o^T e^{-\lambda t}|u_t^\varepsilon|^2 dt = |u_o|^2 + 2 \int_o^T e^{-\lambda t}(f - MB(Z_t^\varepsilon)g(Z_t^\varepsilon), u_t^\varepsilon) dt +$$

$$+ 2 \int_o^T e^{-\lambda t}(MB(Z_t^\varepsilon)u_t^\varepsilon - Mg(Z_t^\varepsilon), B(Z_t^\varepsilon)u_t^\varepsilon - g(Z_t^\varepsilon)) \ dt + Y_\varepsilon$$

where $E(Y_\varepsilon)=0(\varepsilon)$. Let $\Phi(z,u)$ be the solution of the Poisson equation :

$$L \ \Phi \ (z,u)=E(MB(Z_0)u-Mg(Z_0)u,B(Z_0)u-g(Z_0))-$$

$$-(MB(z)u-Mg(z),B(z)u-g(z))+$$

$$+E < MB(Z_0)B(Z_0)u,u > - \ MB(z)B(z)u,u > -$$

$$-E(MB(Z_0)g(Z_0),u)+(MB(z)g(z),u)$$

Apply Ito formula to the process $2\varepsilon^2 e^{-\lambda t}\Phi(Z_t^\varepsilon,u_t^\varepsilon)$, integrate from o to T, and add to the above identity. Taking the expectation in the resulting identity yields :

$$\overline{E}^\varepsilon(X_1)=|u_0|^2+2 \ \overline{E}^\varepsilon \int_0^T e^{-\lambda t}(\overline{f},v_t)dt - e^{-\lambda T}\overline{E}^\varepsilon(|v_T|^2)+0(\varepsilon)$$

Since the mapping $v \to |v_T|^2$ is lower semi-continuous on Ω, we conclude that :

$$\varlimsup_{n \to \infty} \ \overline{E}^{\varepsilon n}(X_1) \leqslant |u_0|^2+2 \ \overline{E} \int_0^T e^{-\lambda t}(\overline{f},v_t)dt - e^{-\lambda T}\overline{E}(|v_T|^2) \quad (3.6)$$

On easily deduce from the caracterization of Q that :

$$\overline{E}(X_1)=|u_0|^2+2 \ \overline{E} \ \int_0^T e^{-\lambda t}(\overline{f},v_t)dt - e^{-\lambda T}\overline{E}(|v_T|^2) \qquad (3.7)$$

It now follows froms $(3.4),(3.5),(3.6)$ and (3.7) :

$$\overline{E}(X) \geqslant 0$$

Let m be any Borel probability measure on $\mathcal{V}$, Φ any non negative bounded continuous functional on $\mathcal{V}$. We have :

$$\int_V \Phi(w)\int_\Omega X(\omega,\omega',w)\overline{Q}(d\omega \ , \ d\omega')m(dw) \geqslant o$$

Using Fubini's theorem and the change of variable $w = v(\omega)- \rho \ \overline{w}, \rho > o$, we deduce :

$$\int_{V \times \overline{\Omega}} \Phi(v-\rho\overline{w})\frac{1}{\rho}X(\omega,\omega',v(\omega)-\rho \overline{w})m_\omega(d\overline{w})\overline{Q}(d\omega,d\omega') \geqslant o$$

where $m_\omega(d\overline{w})$ is a certain transition probability. We can then let $\rho \to o$ by Lebesgue's dominated convergence theorem, yielding :

$$\int_{V \times \overline{\Omega}} \Phi(v(\omega))(\int_0^T e^{-\lambda t}(r_t(\omega')-F(v_t(\omega)),\overline{w}_t)dt)m_\omega(d\overline{w})\overline{Q}(d\omega,d\omega')$$

The result follows from the freedom of choice for Φ and m.
□

CONCLUSION : We have used here the "monotonicity method"
for proving a convergence result for a non linear PDE with
wide-band noise input. One should be able to handle other
types of nonlinearities by means of a compactness argument
(see VIOT [11]), but we don't know what is really possible
in that direction. In any case, non linear terms of the type
$\frac{1}{\epsilon} G(z_t^\epsilon, u_t^\epsilon)$ in equation (3.2) seen to be significantly more
difficult to treat than what we have done here.

E. Pardoux
UER de Mathématiques
Université de Provence
3, Pl. V. Hugo, 13331 Marseille Cedex 3
and Laboratoire de Mécanique et d'Acoustique

REFERENCES

[1] R.N. BHATTACHARYA 'On the Functional Central Limit
Theorem and the law of the Iterated Logarithm for
Markov Processes' *Z. Wahrscheinlichkeitstheorie* $\underline{60}$,
185 - 201 (1982).

[2] G. BLANKENSHIP, G. PAPANICOLAOU 'Stability and Control
of Stochastic Systems with Wide-Band Noise Disturbances
I'. *SIAM J. Appl. Math.* $\underline{34}$, 437 - 476 (1978).

[3] R. BOUC, E. PARDOUX 'Asymptotic Analysis of PDEs with
Wide-Band Noise Disturbances, and Expansion of the
Moments' *Stoch. Anal. and Applic.*, to appear.

[4] H. HUANG, H. KUSHNER 'Weak Convergence and Approximations for Partial Differential Equations with Stochastic
Coefficients. Preprint.

[5] R. KHASMINSKII *Stochastic Stability of Differential
Equations*, Sijthoff and Noordhoff (1980).

[6] H. KUSHNER 'Junp-Diffusion Approximations for Ordinary
      Differential Equations with Wide—Band Random Right
      Hand Side' *SIAM J. Control and Opt.* 17, 729-744 (1979) .

[7] H. KUSHNER, H. HUANG 'Limits for Parabolic Partial
      Differential Equations with Wide—Band Stochastic
      Coefficients' *Stochastics,* to appear.

[8] G. PAPANICOLAOU 'Asymptotic Analysis of Stochastic
      Equations' in *Studies in Probability Theory,*
      M. Rosenblatt Ed., MAA Studies in Math. 18, 111-179
      (1978).

[9] E. PARDOUX 'Equations aux Dérivées Partielles Stochas-
      tiques Nonlinéaires Monotones' Thesis, Univ. Paris XI
      (1975).

[10] A. TOUATI 'Théorèmes de Limite Centrale Fonctionnels
      pour les Processus de Markov' *Ann. Inst. Henri Poincaré,*
      19, 43 - 55 (1983).

[11] M. VIOT 'Solutions Faibles d'Equations aux Dérivées
      Partielles Stochastiques Non linéaires', Thesis,
      Univ. Paris VI (1976).

Hermann Rost

A CENTRAL LIMIT THEOREM FOR A SYSTEM OF
INTERACTING PARTICLES

1. INTRODUCTION

Consider a system of particles moving at random in $\mathbb{Z}^d$ in
such a way that its probability law is stationary in time
and invariant under spatial shifts; denote by $X(j,t)$ the
number of particles at site j at time t, let ρ be its expectation. The corresponding <u>fluctuation processes</u> are the
S'-valued processes N^ε, defined by

(1) $N^\varepsilon(\phi,t) = \varepsilon^{d/2} \cdot \sum_j \phi(\varepsilon j)(X(j,t\varepsilon^{-2}) - \rho)$, $\phi \in S$

(S' : space of Schwartz distributions on R^d,
 S : space of rapidly decaying smooth functions).

The parameter ε ranges between 0 and 1; it measures the ratio of microscopic to macroscopic length unit. In [1] for a
particular class of models it has been shown that

(2) $\lim_{\varepsilon \to 0} EN^\varepsilon(\phi,0)N^\varepsilon(\psi,t) = \chi \cdot \int \phi \cdot T_t \psi \cdot dx$,

where (T_t) is the semigroup on R^d generated by $\kappa/2 \cdot \Delta$, and
χ and κ are explicitly calculable in terms of the given dynamics and the equilibrium density ρ.

As is well known, there exists a Gaussian process N, indexed
by R_+ and S, whose covariance is given by the r.h.s. of (2).
It may also be characterized by the following two properties:

(3a) $EN(\phi,t)^2 = \chi \int \phi^2 \cdot dx$, $\phi \in S$, $t \geq 0$;

(3b)   for each $\phi \in S$, the process $t \to N(\phi,t) - N(\phi,0) - \int_0^t \kappa/2 \cdot N(\Delta\phi,s)ds$
          is a Brownian motion with diffusion constant
          $-u\int \phi \cdot \Delta\phi \cdot dx$ where $u = \chi \cdot \kappa/2$

(see e.g. [2],[3]); we call it the stationary <u>Ornstein-Uh-
lenbeck process</u> with drift operator $\kappa/2 \cdot \Delta$ and <u>driving noise</u>
operator $-u \cdot \Delta$.

We show in this short note that in addition to convergence
243

L. Arnold and P. Kotelenez (eds.), Stochastic Space-Time Models and Limit Theorems, 243–248.
© 1985 by D. Reidel Publishing Company.

of the covariances, as stated in [1], the laws of the proces-
ses N^{ε}, as measures in the Skorokhod space $\mathcal{D}(R_+, S')$, converge
weakly to that of the process N characterized by (3). We show
even more, namely that this convergence already takes place
in the space $\mathcal{D}(R_+, H(-d-4))$. (For notations, see section 2.)
In particular, we thus establish the limiting joint Gaussian
character of the variables $N^{\varepsilon}(\phi, t)$, $t \geq 0$, $\phi \in S$. A similar result, for independent particles, has been proven in [2]; it
is natural that our proof has many arguments in common with
that paper.

2. ASSUMPTIONS AND NOTATIONS

The underlying particle process X is Markovian on a subset
of the space of counting measures on Z^d; its generator is

(4) $Lf(x) = \sum_j c(x_j) \cdot \sum_{|k-j|=1} (f(x+\delta_k - \delta_j) - f(x))$.

The dynamics is determined by the function c which is defined
on the positive integers and is supposed to be increasing and
Lipschitz. (For existence, see e.g. [4].) We consider the
process X only at a fixed equilibrium with distribution at
a single time μ_u, where μ_u is the product measure with identical factors, satisfying

(5) $\mu_u(x_j = m) = Z(u)^{-1} \cdot u^m \cdot \prod_1^m c(n)^{-1}$, $m \in N$.

Here u is a positive parameter characterizing the equilibrium
measure. The following expectations are of interest to us

(6) $\rho := EX(j, t)$

           $\chi := E(X(j, t) - \rho)^2$;

we compute, besides

(7) $u = Ec(X(j, t))$.

In order to fomulate the concept of weak convergence we need
the following notations (as in [3]): $H(m)$, $m \geq 0$, is the completion of the space of infinitely differentiable functions
on R^d with compact support w.r. to the norm

(8) $\|\phi\|_m := (\int \phi \cdot (x^2 - \Delta)^m \phi \cdot dx)^{1/2}$.

The dual of $H(m)$ will be denoted by $H(-m)$. We have the relations

$$S=\bigcap_m H(m), \qquad S'=\bigcup_m H(-m) .$$

We use the symbol ν for elements of the spaces $H(-m)$, the
value of ν at $\phi \in H(m)$ is written as $\nu(\phi)$; random elements in
$H(-m)$ are denoted by N, N^{ε} etc. If we introduce the system
of Hermite functions h_{α}, $\alpha=(\alpha_1,\dots,\alpha_d)$ a multiindex, each
element ϕ of $H(m)$ allows an expansion

(9) $\phi=\sum_{\alpha} c(\alpha) h_{\alpha}$ with $\sum c(\alpha)^2 \cdot (2|\alpha|+d)^m < \infty$

This expansion is meaningful for negative m, too, if for $m<0$
and $\nu \in H(m)$ we interprete the statement

$$\nu=\sum_{\alpha} c(\alpha) h_{\alpha} \quad \text{by} \quad c(\alpha)=\nu(h_{\alpha}) \qquad \text{for all } \alpha.$$

The fluctuation processes N^{ε} have already been defined in (1).
Here we remark only that we can interprete each N^{ε} even as
$H(-m)$-valued process for sufficiently large m (in section 3
we take m=d+4); correspondingly, the evaluation $N^{\varepsilon}(\phi,t)$ in
(1) is meaningful for $\phi \in H(m)$, not only $\phi \in S$.

3. THE CONVERGENCE THEOREM

In this section we prove the following

Theorem. The laws of the processes N^{ε} converge weakly, as $\varepsilon \to 0$,
in the Skorokhod space $D(R_+, H(-d-4))$ to the law of the Ornstein-Uhlenbeck process N defined by (3).

Proof. We begin with an outline of the major steps. We re-
present, for each ϕ, $N^{\varepsilon}(\phi,\cdot)$ as semimartingale and estimate
its drift and martingale part (A). From there, we conclude
that up to any finite time T with high probability the processes N^{ε} can be uniformly approximated in $H(-d-4)$-norm by
a process with values in a finite dimensional space (B).
Hence tightness of the family N^{ε} is reduced to tightness of
$N^{\varepsilon}(\phi,\cdot)$ for each ϕ separately (C). Finally, using the result
established in [1], we show that any limiting process satis-
fies the conditions (3), which proves the desired convergence
statement.

(A) From the definition of the dynamics it follows that for
each ϕ the processes $M^{\varepsilon}_i(\phi,\cdot)$, i=1,2, are martingales:

(10a) $M_1^\varepsilon(\phi,t):=N^\varepsilon(\phi,t)-N^\varepsilon(\phi,0)-\int_0^t A^\varepsilon(\phi,s)\,ds,$

where

(10b) $A^\varepsilon(\phi,t)=\varepsilon^{d/2}\sum_j\left[c(X(j,t\varepsilon^{-2}))-u\right]\cdot\Delta^\varepsilon\phi(\varepsilon j)$

$\qquad(\Delta^\varepsilon\phi(y):=\sum_{|e|=1}\varepsilon^{-2}\cdot(\phi(y+\varepsilon\cdot e)-\phi(y)));$

(11a) $M_2^\varepsilon(\phi,t):=(M_1^\varepsilon(\phi,t))^2-\int_0^t B^\varepsilon(\phi,s)\,ds\ ,$

where

(11b) $B^\varepsilon(\phi,t):=\varepsilon^d\cdot\sum_j c(X(j,t\varepsilon^{-2}))\cdot\sum_{|k-j|=1}\varepsilon^{-2}\cdot(\phi(\varepsilon k)-\phi(\varepsilon j))^2\ ;$

further , the jump size of $M_1^\varepsilon(\phi,\cdot)$ is bounded by

$\qquad\varepsilon^{d/2+1}\sup_y|\nabla\phi(y)|\ .$

We have the estimate for T fixed:

(12a) $E\sup_{t\le T}(\int_0^t A^\varepsilon(\phi,s)\,ds)^2$

$\qquad\le E(\int_0^T|A^\varepsilon(\phi,s)|\,ds)^2\le T^2\cdot E(A^\varepsilon(\phi,0))^2\ ,$

by stationarity. An elementary calculation shows that the right hand side is bounded by

(12b) $T^2\cdot\mathrm{var}\ c(X(0,0))\cdot\left\{\|\phi\|_2^2+\varepsilon^2\cdot K_1\|\phi\|_3^2\right\}$

with K_1 depending only on d. The variance of $c(X(0,0))$ is finite, as follows easily from (5) and the Lipschitz property of $c(\cdot)$. Similarly

(13) $E(M_1^\varepsilon(\phi,T))^2=T\cdot EB^\varepsilon(\phi,0)=$

$\qquad\le T\cdot Ec\cdot(\|\phi\|_1^2+K_2\cdot\varepsilon^2\cdot\|\phi\|_2^2)\ .$

B.   From the preceding estimate follows the existence of a constant K(T) such that, uniformly in $0<\varepsilon<1$,

(14) $E\sup_{t\le T}(N^\varepsilon(\phi,t))^2\le K(T)\cdot\|\phi\|_3^2\ .$

We choose ϕ equal to the Hermite function h_α and get

(15) $E \sup_{t \leq T} (N^\varepsilon(h_\alpha,t))^2 \leq K(T) \cdot (2|\alpha|+d)^3$

If we multiply (15) by $(2|\alpha|+d)^{-d-4}$ and sum over α, the r.h.s. converges; hence, in the norm of $H(-d-4)$, the processes N^ε are, uniformly in ε, approximated by their projections on a finite dimensional subspace.

C. In order to show tightness for fixed $\phi \in S$ of the family $N^\varepsilon(\phi,\cdot)$, we do it separately for $M_1^\varepsilon(\phi,\cdot)$ and $\int A^\varepsilon(\phi)ds$. The second family is tight, because as in (12a)

(16) $E(\int_r^t A^\varepsilon(\phi,s)ds)^2 \leq (t-r)^2 \cdot K(\phi)$,

which implies tightness even in $C(R_+,R)$ by the Kolmogorov criterion.
The martingales $M_1^\varepsilon(\phi)$ are convergent to Brownian motion, hence tight; indeed, $B^\varepsilon(\phi,s)$ converges in L^2 to $u \cdot \|\phi\|_1^2$, therefore the integrals $\int_0^t B^\varepsilon(\phi,s)ds$ converge to $t \cdot u \|\phi\|_1^2$. Since the maximal jump size of $M_1^\varepsilon(\phi)$ tends to zero (A), the claim follows ([5]).

D. By th. 3 in [1], we have

(17) $\lim_{\varepsilon \to 0} \int_0^t (A^\varepsilon(\phi,s) - u/\chi \cdot N^\varepsilon(\Delta\phi,s))ds = 0$

in L^2. (In this argument the specific properties of the model have been used; it does not follow from pure Sobolev space reasoning.) Hence any limit process N of a sequence $N_{\varepsilon'}$, $\varepsilon' \to 0$, must satisfy, because of (10), (11), and (C),

(18) $N(\phi,t) - N(\phi,0) - \int_0^t u/\chi N(\Delta\phi,s)ds$

        is Brownian motion with diffusion constant

$u \cdot \|\phi\|_1^2$

Since obviously the relation

(19) $\lim_{\varepsilon \to 0} EN^\varepsilon(\phi,t)^2 = \chi \cdot \int \phi^2 \cdot dy$

holds, any limit process N is identified as the stationary Ornstein-Uhlenbeck process described in (3). This proves the theorem.

H. Rost
Institut für Ang. Math.
Im Neuenheimer Feld 294
6900   Heidelberg
West Germany

REFERENCES

1   Th. Brox, H. Rost: 'Equilibrium fluctuations of stocha-
     stic particle systems: the role of conserved quantities.'
     Ann.Probability 12 (1984),

2   A. Martin-Löf: 'Limit theorems for the motion of a Pois-
     son system of independent Markovian particles at high
     density.' Z.Wahrscheinlichkeitstheorie verw. Geb. 34
     (1976), 205-223.

3   R. Holley, D. Stroock: 'Generalized Ornstein-Uhlenbeck
     processes and infinite particle Brownian motion.' Publ.
     RIMS Kyoto Univ. 14 (1978), 741-788.

4   Th. Liggett: 'An infinite particle system with zero range
     interaction.' Ann.Probability 1 (1973), 240-253.

5   R.S. Liptser, A.N. Shiryayev: 'A functional central li-
     mit theorem for semimartingales.' Th. Probability Appl.
     25 (1980), 667-688.

Hans Zessin

MOMENTS OF STATES OVER NUCLEAR LSF SPACES

Introduction: The purpose of the present paper is to investigate the problem of weak convergence of a sequence of states over nuclear LSF spaces V (including $S(\mathbb{R}^\nu)$ and $\mathcal{D}(\mathbb{R}^\nu)$) from assumptions on the asymptotic behaviour of their moments; here by a state over V we understand a continuous linear random function over V. This paper is closely related to [16] where the same problem has been studied for random measures; it is also closely related to the papers of Fernique [4] as well as Dobrushin/Minlos [3], where the basic tools on linear random functions are developed.

The main results, a uniqueness and continuity theorem for the moments of states, can be found in § 1: In theorem 1.1 Carleman's one-dimensional uniqueness condition (1.2) assures uniqueness of a linear process Z in terms of its moments $(\nu_Z^{2n})_n$. This condition is: $\sum_{n\geq 1} \nu_Z^{2n}(f,\overset{2n}{\cdots},f)^{-1/2n}$ diverges for each $f \in V$. The continuity theorem 1.3 establishes weak convergence of a sequence of states (P_n) to some limiting state under the assumptions that for each k the k-th moments of P_n converge to some limit, that these limit moments satisfy Carleman's more-dimensional uniqueness criterium (1.8), and finally under condition (1.9) resp. (1.9') that yields relative compactness of $\{P_n\}$. Theorem 1.2 gives a partial converse of theorem 1.3.

In § 2 we consider the special case $V = S(\mathbb{R}^\nu)$ and give a sufficient condition for relative compactness of $\{P_n\}$ in terms of the second moments, namely: $\sup_n \nu_{P_n}^2(f,f) \leq C \cdot \|f\|_m^2$ for each $f \in S(\mathbb{R}^\nu)$ for some $C > 0$ and $S(\mathbb{R}^\nu)$ - norm $\|\cdot\|_m$. We then comment on the results obtained in connection with the existence problem of the φ_ν^4 - model of Euclidean quantum field theory.

Acknowledgement: I am grateful to M. Röckner for useful discussions.

L. Arnold and P. Kotelenez (eds.), Stochastic Space-Time Models and Limit Theorems, 249–261.
© 1985 by D. Reidel Publishing Company.

§ 1 A Uniqueness and Continuity Theorem for Moments of States over nuclear LSF spaces

Let V be a linear topological space and $V*$ its dual equipped with weak topology. V^*_{alg} denotes the algebraic dual of V. For $f \in V$, $\mu \in V^*_{alg}$ define $\xi_f(\mu) = \mu(f)$. Let $B_0(V^*_{alg}) = \sigma(\xi_f; f \in V)$ be the smallest σ-algebra of subsets in V^*_{alg} with respect to which all the functions ξ_f, $f \in V$, are measurable. For $f \in V$, $\mu \in V*$ we set $\zeta_f(\mu) = \mu(f)$ and denote by $B_0(V*) = \sigma(\zeta_f; f \in V)$ the σ-algebra in $V*$ generated by the functions ζ_f, $f \in V$.

Definition: A linear map Z from V into the set of random variables over a probability space (Ω, F, P) is called a linear process over V (on (Ω, F, P)). Given a probability measure P on $(V*, B_0(V*))$ the linear process Z over V on $(V*, B_0(V*), P)$ defined by $Z : f \to \zeta_f$ is called a state over V (given by P). We also call a probability measure P on $(V*, B_0(V*))$ a state over V. A sequence (P_n) of states over V converges weakly to a state P, and we write $P_n \Rightarrow P$, if $P_n(\varphi) \to P(\varphi)$ for each bounded continuous real function φ on $V*$. Two linear processes Z_1, Z_2 over V on probability spaces (Ω_1, F_1, P_1), (Ω_2, F_2, P_2) are said to be (probabilistically) equivalent, if for all $k \in \mathbb{N}$, $f_1, \ldots, f_k \in V$ the corresponding k-dimensional distributions coincide, i.e.

$$P_{1,(Z_1(f_1),\ldots,Z_1(f_k))} = P_{2,(Z_2(f_1),\ldots,Z_2(f_k))} .$$

A linear process Z over V on (Ω, F, P) is called of k-th order, if for each $f_1, \ldots, f_k \in V$

$$(1.1) \qquad \nu_Z^k(f_1, \ldots, f_k) = \int_\Omega Z(f_1) \cdot \ldots \cdot Z(f_k) \ dP$$

exists and is finite. In this case ν_Z^k is called the moment of Z of k-th order. For a state P over V we write also ν_P^k. Z is called of infinite order if it has moments of all orders.

By Hölder's inequality a sufficient condition for a linear process Z over V to be of k-th order is

$\int_\Omega |Z_f|^k \ dP < +\infty$ for each $f \in V$.

Theorem 1.1: Let Z be a linear process over V on $\overline{(\Omega,F,P)}$ of infinite order. Suppose that

$$(1.2) \qquad \sum_{n=1}^{\infty} \nu_Z^{2n} \, (f,\overset{2n}{...},f)^{-1/2n} = +\infty \quad \text{for each} \quad f \in V ,$$

then Z is uniquely determined by its moments up to equivalence.

Proof: It is well known that for two linear processes over V to be equivalent it is necessary and sufficient that their one-dimensional distributions coincide. But (1.2) just means that the one-dimensional distributions of Z satisfy Carleman's criterium (see [16], Lemma 1.4) and therefore are uniquely determined by its moments.

Example 1.1: One important example of a linear process over V is the Gaussian process. This is a linear process Z over V s.th. for each $f \in V$ the random variable $Z(f)$ is Gaussian. It is well known that its moments are

$$(1.3) \qquad \nu_Z^k \, (f_1,\ldots,f_k) = \sum_{J=\{J_1,\ldots,J_\ell\}} \prod_{i=1}^{\ell} P(\prod_{j \in J_i} Z(f_j)) ,$$

where J is summed over all pair partitions of $\{1,\ldots,k\}$. By a pair partition we mean a partition $\{J_1,\ldots,J_\ell\}$ of $\{1,\ldots,k\}$ into disjoint subsets J_i , s.th., in the case when k is even, each J_i has two elements and in the case when k is odd, exactly one J_i has a single element. The rest has two elements.

$$(1.4) \qquad \nu_Z^{2n} \, (f,\overset{2n}{...},f) = \frac{(2n)!}{2^n \cdot n!} \cdot \nu_Z^2 \, (f,f)^n , \quad f \in V ,$$

which immediately implies that condition (1.2) is satisfied, so that by theorem 1.1 Z is uniquely determined (up to equivalence) by its moments and thus, on account of (1.3), by its first and second moment.

Theorem 1.2: Let (P_n) be a sequence of states over V of infinite order. If P_n converges weakly to some state P and if

$$(1.5) \qquad \sup_n \nu_{P_n}^k \, (f,\overset{k}{...},f) < +\infty , \quad k \in \mathbb{N} , \ f \in V .$$

then P is of infinite order and $v^k_{P_n}(f_1,\ldots,f_k) \vec{\underset{n}{\to}}$
$v^k_P(f_1,\ldots,f_k)$, $k \in \mathbb{N}$, $f_1,\ldots,f_k \in V$.

The proof is left to the reader.

In the following we assume that V is nuclear and an LSF-space, i.e. the strict inductive limit of a sequence of separable Frechêt spaces. Note that the space $S(\mathbb{R}^v)$ is a nuclear separable Frechêt space; the space $D(\mathbb{R}^v)$ is a nuclear LSF space (see [13]). Recall that in a nuclear space V the topology can be generated by a family $(q_\alpha)_{\alpha \in A}$ of seminorms, each of which originates from a positive semi-definite Hermitian form on $V \times V$ (see [13]). Recall also that $B_0(V*)$ coincides with the σ-algebra $B(V*)$ of Borel subsets of $V*$ if V is an LSF space (see [3], Prop. 1, e.g.).

We denote by $(q^*_\alpha)_{\alpha \in A}$ the family of seminorms on $V*$ defined by

(1.6) $q^*_\alpha(\mu) = \sup \{ |\mu(f)| : q_\alpha(f) \le 1 \}$, $\mu \in V*$.

Theorem 1.3: Let V be a nuclear LSF space and let (P_n) be a sequence of states over V , each of which is of infinite order. Suppose that

(1.7)   for each $k \in \mathbb{N}$, $f_1,\ldots,f_k \in V$ the limits
$v^k(f_1,\ldots,f_k) = \lim_{n\to\infty} v^k_{P_n}(f_1,\ldots,f_k)$ exist and

(1.8) $\displaystyle\sum_{n=1}^{\infty} [\sum_{i=1}^{k} v^{2n}(f_i,\overset{2n}{\ldots},f_i)]^{-1/2n} = +\infty$, $k \in \mathbb{N}$,

$f_1,\ldots,f_k \in V$.

If finally

(1.9)   there exists $\alpha \in A$ s.th.

$\lim_{N\to\infty} \sup_n P_n \{q^*_\alpha > N\} = 0$ or equivalently

(1.9')   for each $\varepsilon > 0$ and $a > 0$ there exists a
           neighborhood U of 0 in V s.th.

$$\sup_n P_n \{\sup_{f \in U} |\zeta_f| > a\} < \varepsilon \quad ,$$

then there exists a unique state P over V having moments
of all orders s.th. $P_n \Rightarrow P$ and $v_P^k = v^k$ for each $k \in \mathbb{N}$.
Moreover P is k-continuous, i.e. the mapping $f \to \zeta_f$
is continuous with respect to the topologies in V and
$L_k(P)$, $k \in \mathbb{N}$.

Proof: 1. We first prove the equivalence of (1.9) and (1.9').
$\overline{(1.9)} \Rightarrow (1.9')$: Let $a,\varepsilon,\delta > 0$ and set $U_{\alpha,\delta} = \{f \in V :$
$q_\alpha(f) < \delta\}$. Then $P_n \{\sup_{f \in U_{\alpha,\delta}} |\zeta_f| > a\} \leq P_n \{q_\alpha^* > \frac{a}{\delta}\}$.
Choosing δ in such a way that $\sup_n P_n \{q_\alpha^* > \frac{a}{\delta}\} < \varepsilon$,
yields (1.9').

$(1.9') \Rightarrow (1.9)$: For $a,\varepsilon > 0$ choose a neighborhood U of
0 in V s.th. $\sup_n P_n \{\sup_{f \in U} |\zeta_f| > a\} < \varepsilon$. Now there exists
α and N s.th. $\frac{a}{N} \{q_\alpha \leq 1\} \subseteq U$. Then
$$\sup_n P_n \{q_\alpha^* > N\} \leq \sup_n P_n \{\sup_{f \in U} |\zeta_f| > a\} < \varepsilon \quad .$$

2. Let $k \in \mathbb{N}$, $r_1,\ldots,r_k \in \mathbb{N}_0$, $f_1,\ldots,f_k \in V$. Combining
assumption (1.7) with lemma 1.4 of [16] yields the existence
of a probability measure $\mu_{f_1,\ldots,f_k}$ on $\mathbb{R}^k$ which is a
solution of the $(\mathbb{R}^k)$ moment problem corresponding to the
moments

$$\alpha_{f_1,\ldots,f_k}(r_1,\ldots,r_k) = v^{r_1+\ldots+r_k}(\overbrace{f_1,\ldots,f_1}^{r_1},\ldots,\overbrace{f_k,\ldots,f_k}^{r_k}),$$

because

$$\sum_{r_1=0}^{N_1} \ldots \sum_{r_k=0}^{N_k} a_{r_1,\ldots,r_k} \cdot \alpha_{f_1,\ldots,f_k}(r_1,\ldots,r_k) =$$

$$= \lim_{n \to \infty} \int_{\mathbb{R}^k} \sum_{r_1=0}^{N_1} \ldots \sum_{r_k=0}^{N_k} a_{r_1,\ldots,r_k} \cdot x_1^{r_1}\ldots x_k^{r_k} \, P_{n,(\zeta_{f_1},\ldots,\zeta_{f_k})}(dx)$$

is nonnegative if the polynomial in the last integral is nonnegative on $\mathbb{R}^k$. Thus

(1.10) $\nu^{r_1+\ldots+r_k}(f_1,\ldots,f_1,\ldots,f_k,\ldots,f_k) =$

$$= \int_{\mathbb{R}^k} x_1^{r_1}\ldots x_k^{r_k}\ \mu_{f_1,\ldots,f_k}(dx)$$

for each $r_1,\ldots,r_k \in \mathbb{N}_0$.

3. From Carleman's more-dimensional uniqueness criterium (see lemma 1.4 of [16], e.g.) we know that the probability measures $\mu_{f_1,\ldots,f_\alpha}$ are uniquely determined by ist moments on account of assumption

(1.8) $\displaystyle \sum_{n\geq 1} (\sum_{i=1}^{k} \nu^{2n}(f_i,\ldots,f_i))^{-1/2n} = +\infty$.

Therefore

(1.11) $P_{n,(\zeta_{f_1},\ldots,\zeta_{f_k})} \Rightarrow \mu_{f_1,\ldots,f_k}$, $f_1,\ldots,f_k \in V$,

follows from (1.7) combined with (1.10) by lemma 1.5 in [16].

4. The next step is to construct a probability measure P on $B_0(V^*_{alg})$ having the $\mu_{f_1,\ldots,f_k}$ as marginal distributions: Let F be a k-dimensional subspace of V and $f_1,\ldots,f_k$ a basis of F . Define a probability measure P_F on $(V^*_{alg},(\zeta_{f_1},\ldots,\zeta_{f_k})^{-1}(B_k))$ by

(1.12) $P_F((\zeta_{f_1},\ldots,\zeta_{f_k})^{-1}(A)) = \mu_{f_1,\ldots,f_k}(A)$, $A \in B_k$.

Here B_k denotes the σ - algebra of Borel subsets in $\mathbb{R}^k$. P_F is well defined, since $(\zeta_{f_1},\ldots,\zeta_{f_k})$ is surjective. Moreover (1.11) immediately implies that the family $\{P_F$; F finite-dimensional subspace of $V\}$ is consistent. Therefore we can define an additive probability measure P on the Boolean algebra $\tilde{B} = \cup_F \sigma(\xi_f ; f \in F)$ in V^*_{alg} by

(1.13) $P(B) = P_F(B)$, $B \in \sigma(\xi_f ; f \in F)$.

Now it is well known that P is σ-additive on $\tilde{B}$. (Le-
nard's theorem; see [12] e.g.) Therefore P can be extended to a probability measure P on $\sigma(B) = B_0(V_{alg}^*)$. We
note also that the linear process Z over V on
$(V_{alg}^*, B_0(V_{alg}^*), P)$ given by $Z : f \to \xi_f$ has moments of all
orders s.th. $v_Z^k = v^k$, $k \in \mathbb{N}$. This follows immediately
from (1.10).

5. We now show that P in continuous, i.e. $f_\gamma \to f$ in V
implies $\xi_{f_\gamma} \to \xi_f$ in measure, i.e. $P\{|\xi_{f_\gamma}| > a\} \to 0$ for
each $a > 0$. Since for each $f \in V , P_{n,\zeta_f} \Rightarrow P_{\xi_f}$ we have
for each $a > 0$ and $f \in V$

$$P\{|\xi_f| > a\} \leq \lim_{n\to\infty} \inf P_n \{|\zeta_f| > a\}$$

$$\leq \sup_n P_n \{\sup_{f\in U}|\zeta_f| > a\}$$

for each neighborhood U of 0 in V containing f .
Continuity of P thus follows from (1.9'). Therefore by
Minlos' theorem the nuclearity of V implies that Z is
equivalent to a state over V which we again denote by P
(see [3] or [4]).

6. Finally we show that $P_n \Rightarrow P$. As we know that

$$P_{n,(\zeta_{f_1},\ldots,\zeta_{f_k})} \Rightarrow P_{(\zeta_{f_1},\ldots,\zeta_{f_k})} \quad , \quad (k \in \mathbb{N} , f_1,\ldots,f_k \in V),$$

it suffices to show that $\{P_n\}$ is relatively compact with
respect to weak topology. For this it is sufficient that
$\{P_n\}$ is tight, i.e. for each $\varepsilon > 0$ there is a compact
subset K of V^* s.th. $P_n(K) \geq 1 - \varepsilon$ for each n .
This follows from Prohorov's theorem which is valid in our
situation, since V^* is the weak dual of an LSF space and
thus a regular, standard Borel space (see [4] or [14]).
We construct K in the following way: By the Banach-Stein-
haus theorem a subset $X \subset V^*$ is weakly precompact if and
only if it is equicontinuous, i.e. given $\varepsilon > 0$ there

exists a neighborhood U of 0 in V s.th.
$\sup_{f \in U} \sup_{\mu \in X} |\mu(f)| \leq \varepsilon$. But by assumption (1.9') for $\varepsilon > 0$,
$k \in \mathbb{N}$ there exists a neighborhood U_k of 0 in V s.th.
for each n $P_n \{ \sup_{f \in U_k} |\zeta_f| > \frac{1}{k} \} \leq \frac{\varepsilon}{2^k}$. Thus

$P_n(\cap_{k \geq 1} \{ \sup_{f \in U_k} |\zeta_f| \leq \frac{1}{k} \}) \geq 1 - \varepsilon$ for each n . Set
$K_\varepsilon = \overline{\cap_{k \geq 1} \{ \sup_{f \in U_k} |\zeta_f| \leq \frac{1}{k} \}}$. K_ε is equicontinuous and therefore weakly precompact. Thus $K = \overline{K_\varepsilon}$ is compact in V^* and
has the required properties.

7. Continuity for each k follows at once from prop. 4 in
[3] because P is of infinite order. This completes the
proof of the theorem.

We now discuss the problem of weak convergence to a Gaussian
process P .

Corollary 1.1: Let (P_n) be a sequence of states over a
nuclear LSF space V of infinite order satisfying (1.7)
and (1.9) resp. (1.9'). If furthermore the limiting moments
ν^n , $n \in \mathbb{N}$, appearing in (1.7) obey (1.3), then there
exists a unique Gaussian state P over V with mean value
ν^1 and sec.moment ν^2 s.th. $P_n \Rightarrow P$. Moreover, P is
k-continuous for each $k \in \mathbb{N}$.

Proof: 1. Note that $(\nu^{2n})_n$ satisfies the uniqueness condition (1.8) on account of (1.4). Thus we are in the situation of theorem 1.3. That the limiting state P is
Gaussian with mean ν^1 and sec.moment ν^2 follows immediately from theorem 1.1 .

§ 2 The case $V = S(\mathbb{R}^\nu)$, $\nu \geq 1$

We first summarize some facts about Schwartz distributions:
(cf. [8] and [9]): The inner product and the norm in
$L_2(\mathbb{R}^\nu)$ are denoted by $(,)$ and $\| \|$ respectively.
Given $n \in \mathbb{N}_0 = \mathbb{N} \cup \{0\}$ the Hermite function of order n
corresponding to the Hermite polynomial H_n of order n
is given by

(2.1) $h_n(x) = c_n \cdot H_n \cdot \exp(-\frac{x^2}{2})$, $x \in \mathbb{R}$,

where c_n is a constant for which $\|h_n\| = 1$. It is well
known that $\{h_n ; n \geq 0\}$ is an orthonormal base in $L_2(\mathbb{R})$.
Define for $\beta = (\beta_1, \ldots, \beta_\nu) \in \mathbb{N}_0^\nu$ the function $h_\beta : \mathbb{R}^\nu \to \mathbb{R}^\nu$
by

(2.2) $h_\beta(x_1, \ldots, x_\nu) = h_{\beta_1}(x_1) \cdot \ldots \cdot h_{\beta_\nu}(x_\nu)$.

$\{h_\beta ; \beta \in \mathbb{N}_0\}$ is an orthonormal base in $L_2(\mathbb{R}^\nu)$. The m-norm
$\| \ \|_m$ is $S(\mathbb{R}^\nu)$ is defined as follows:

(2.3) $\|f\|_m^2 = \sum_\beta (f, h_\beta)^2 \cdot (2|\beta| + \nu)^m$, $f \in S(\mathbb{R}^\nu)$, $m \in \mathbb{N}_0$,

where $|\beta| = \beta_1 + \ldots + \beta_\nu$. $S(\mathbb{R}^\nu)$ is topologized by the set
of norms $\| \ \|_m$, $m \in \mathbb{N}_0$, and is nuclear as such. It is
a real pre-Hilbert space with an inner product corresponding
to $\| \ \|_m$. The $(-m)$ – norm $\| \ \|_{-m}$ in $S^*(\mathbb{R}^\nu)$ is defined
as follows:

(2.4) $\|\mu\|_{-m}^2 = \sum_\beta \mu(h_\beta)^2 \dfrac{1}{(2|\beta| + \nu)^m}$, $\mu \in S^*(\mathbb{R}^\nu)$.

It is easy to see that for each $m \in \mathbb{N}_0$, $\mu \in S^*(\mathbb{R}^\nu)$

(2.5) $\|\mu\|_{-m} = \sup \{|\mu(f)| : f \in S(\mathbb{R}^\nu) , \|f\|_m \leq 1\}$.

The definition of $\| \ \|_{-m}$ suggests a sufficient condition
for (1.9) in terms of the second moments. Indeed we have

Theorem 2.1: Let (P_n) be a sequence of states over
$S(\mathbb{R}^\nu)$, $\nu \geq 1$. If

(2.6)   for each $k \in \mathbb{N}$ there exists $C > 0$ and
           $m \in \mathbb{N}_0$ s.th.

           $\sup_n \nu_{P_n}^k (f, \overset{k}{\ldots}, f) \leq C \cdot \|f\|_m^k$, $f \in S(\mathbb{R}^\nu)$,

then there exists a state P over $S(\mathbb{R}^\nu)$ of infinite order
and a subsequence $(P_{n_\ell})_\ell$ s.th. $P_{n_\ell} \Rightarrow P$ and

$$\nu^k_{P_{n_\ell}} (f_1,\ldots,f_k) \to \nu^k_{P_\ell} (f_1,\ldots,f_k) \;,\; k \in \mathbb{N} \;,\; f_1,\ldots,f_k \in S(\mathbb{R}^\nu).$$

Conversely, let each P_n be of infinite order satisfying (1.7) and (1.8). If furthermore

(2.7)   there exists $C > 0$ and $m \in \mathbb{N}_0$ s.th.

$$\sup_n \nu^2_{P_n} (f,f) \leq C \cdot \|f\|^2_m \;,\; f \in S(\mathbb{R}^\nu) \;,$$

then there exists a unique state P over $S(\mathbb{R}^\nu)$, k-con-
tinuous for each k , of infinite order s.th. $P_n \Rightarrow P$ and
$\nu^k_P = \nu^k$, $k \in \mathbb{N}$.

Proof: 1. We first treat the second part of the theorem.
It is easy to see that there exists $m' \geq m$ s.th.

$$\Sigma_\beta \; \|f_\beta\|^2_m < + \infty \;,\; \text{where} \; f_\beta = \frac{h_\beta}{(2|\beta| + \nu)^{m'/2}} \;.$$

Therefore on account of (2.7)

$$\Sigma_\beta \sup_n \nu^2_{P_n} (f_\beta,f_\beta) < + \infty \;.$$

But this in turn implies (1.9) because

$$\sup_n P_n \{\| \; \|_{-m'} > N\} \leq \frac{1}{N^2} \sup_n P_n(\| \; \|^2_{-m'})$$

$$\leq \frac{1}{N^2} \cdot \Sigma_\beta \sup_n \nu^2_{P_n} (f_\beta,f_\beta) \xrightarrow[N \to \infty]{} 0 \;.$$

Thus theorem 1.3 applies.

2. Note that (2.6) implies (2.7) and therefore via (1.9)
relative compactness of $\{P_n\}$. Thus there exists a state
P over $S(\mathbb{R}^\nu)$ and a subsequence (P_{n_ℓ}) s.th. $P_{n_\ell} \Rightarrow P$.
Now theorem 1.2 applies and proves the first part.

Combining the second part of theorem 2.1 with Corollary 1.1
we have

Corollary 2.1: Let (P_n) be a sequence of states over
$S(\mathbb{R}^\nu)$, $\nu \geq 1$, of infinite order satisfying (1.7) and (2.7).
If furthermore the limiting moments ν^n , $n \in \mathbb{N}$, appearing
in (1.7) obey (1.3) then there exists a unique Gaussian

state P over $S(\mathbb{R}^\nu)$ with mean value ν^1 and sec.moment ν^2 s.th. $P_n \Rightarrow P$. Moreover, P is k-continuous for each $k \in \mathbb{N}$.

Remarks: (1) Condition (2.7) is fullfilled for the socalled φ_ν^4-lattice field models P_n of Euclidean quantum field theory. The limit of the sequence (P_n) considered there is the socalled continuum resp. scaling limit. For the details we refer to [2], [11] and [15]. The main problem of Euclidean quantum field theory is twofold: (α) to esta- blish the existence of the continuum resp. scaling limit P of the sequence (P_n) and (β) to show that P is non- trivial, i.e. non-Gaussian. Here the second part of the problem is the much more difficult and is not answered yet for dimension $\nu = 4$.

The following remarks should show that theorem 2.1 combined with recent results of Palmer/Tracy [11] give a complete answer to the existence problem for dimension $\nu = 2$ (in. the case of Ising-models instead of φ_2^4-models), and, combined with a result of Glimm/Jaffe [7]; a not completely satisfactory answer in the other cases.

(2) For a sequence of two-dimensional Ising-models (P_n) Palmer/Tracy [11] verified conditions (1.7) and (1.8). Since (2.7) is also satisfied, by the second part of theorem 2.1 the scaling limit P of (P_n) exists and is uniquely determined. This limit is non-Gaussian.

(3) Consider a sequence (P_n) of φ_ν^4-models, $\nu \geq 1$. We have remarked that (2.7) is true. Now by socalled Gaussian domination (see Newman [10]) condition (2.7) implies bounds for the moments $\nu_n^k (f,\ldots,f)$, $k \in \mathbb{N}$, $f \in S(\mathbb{R}^\nu)$, uni- formly in n, in such a way that (2.6) holds. This has been remarked by Glimm/Jaffe [7]. Thus the first part of theorem 2.1 applies. To summarize: In a sequence (P_n) of φ_ν^4-lattice field models with arbitrary $\nu \geq 1$ one can find a sequence $(P_{n_\ell})_\ell$ which in the scaling limit con- verges weakly to some state P over $S(\mathbb{R}^\nu)$ of infinite order s.th. the socalled Schwinger functions $\nu_{n_\ell}^k (f_1,\ldots,f_k)$ converge to the corresponding Schwinger function ℓ of P. This improves the result of Glimm and Jaffe. For dimension $\nu = 2,3$ it has been shown that P is non-Gaussian. (For a recent elementary proof of this fact see [6]). As has

been remarked by Fröhlich et al. [6] this result is non-satisfactory: "Aside from its inherent non-constructiveness, certain natural and desirable properties (such as uniqueness of the limit) go unestablished."

(4) We finally remark that Aizenman [1] and Fröhlich [5] have shown that a sequence of φ^4_ν lattice field models satisfies the assumptions of Corollary 2.1 if $\nu > 4$. Thus in this case the weak scaling resp. continuum limit P is Gaussian.

Hans Zessin
Universität Bielefeld
Fakultät für Mathematik
Postfach 86 40
4800 Bielefeld 1
Federal Republik of Germany

References:

[1] Aizenman, M.: 1982, 'Geometric analysis of φ^4 fields and Ising models', Commun. math. Phys. 86, 1-48.
[2] Brydges, D.: 1982, 'Field theories and Symanzik's polymer representation', in: Gauge theories: fundamental interactions and rigoros results. Dita, P., Georgescu, V., Purice, R., eds. Boston-Basel-Stuttgart: Birkhäuser.
[3] Dobrushin, R.L., Minlos, R.A.: 'The moments and polynomials of a generalized random field'. Theory of probability and its applications 23, (1978), 686-699.
[4] Fernique, X.: 1967, 'Processes linéaires, processes généralisés'. Ann. Inst. Fourier 17, 1-92.
[5] Fröhlich, J.: 1982, 'On the triviality of $\lambda\phi^4_d$ theories and the approach to the critical point in $d_{(\geq)} 4$ dimensions'. Nucl. Phys. B200, 281-296.
[6] Fröhlich, J., Brydges, D.C., Sokal, A.D.: 1983, 'A new proof of the existence and nontriviality of the continuum ϕ^4_2 and ϕ^4_3 quantum field theories'. Preprint.
[7] Glimm, J., Jaffe, A.: 1974, 'Remark on the existence of φ^4_4'. Phys. review letters 33, 440-442.

[8] Holley, R.A., Stroock, D.W.: 1978, 'Generalized Ornstein-
    Uhlenbeck processes and infinite particle branching
    Brownian motions'. Publ. RIMS, Kyoto Univ., 14, 741-788.
[9] Ito, K.: 1983, 'Distribution-valued processes arising from
    independent Brownian motions'. Math. Z. 182, 17-33.
[10] Newman, C.M.: 1975, 'Gaussian correlation inequalities for
    ferromagnets'. Z. Wahrscheinlichkeitstheorie verw. Gebiete
    33, 75-93.
[11] Palmer, J., Tracey, C.: 1981, 'Two-dimensional Ising corre-
    lations: Convergence of the scaling limit'. Adv. appl.
    math. 2, 329-388.
[12] Reed, M.C.: 1973, 'Functional analysis and probability theory'.
    In: Constructive quantum field theory. Velo. G., Wightman,
    A. eds.. Berlin-Heidelberg-New-York: Springer.
[13] Schäfer, H.H.: 1971, 'Topological vector spaces'. New York:
    Springer.
[14] Smolyanov, O.G., Fomin, S.V.: 1976, 'Measures on topological
    linear spaces', Uspekki Matem. Nauk 31, 3-56.
[15] Sokal, A.D.: 1982, 'An alternate constructive approach to
    the φ_3^4 quantum field theory, and a possible destructive
    approach to φ_4^4'. Ann. Inst. Henn Poincaré A 37, 317-398.
[16] Zessin, H.: 1983, 'The method of moments for random measures'.
    Z. Wahrscheinlichkeitstheorie verw. Gebiete 62, 395-409.

Subject Index

Abstract Wiener space   2
additive process on nuclear space   176
approximation, Gaussian   7
approximation, non-Gaussian   7

Banach space valued process   53
Banach valued Sobolev space   7
bifurcation point   5
boundary noise   81, 89

Canonical form of semimartingale   195
central limit theorem   4, 180, 243
coloured noise   141
complex   26
continuity of sample paths   87, 171
contraction type semigroup   97
cosurface   26
critical expansion   183
critical exponent   185, 186
cylindrical Brownian motion   2, 131
- distribution   153
- function   56, 153
- process   144, 146, 148
- process, generalized   144
- random variable   143, 145
- test function   153

Delay-differential equation, stochastic   95
Dirichlet forms in infinite dimension   11, 14

Evolution equation, stochastic   95, 98, 114
- ,stochastic, linear   114
- ,stochastic, nonlinear   125
extension of operators   2
external noise   2, 7

Fluctuations   1, 179, 183, 243
free quantum field   20
Fubini theorem, stochastic   108
functional limit theorem   191

Gaussian white noise   144